AF732481

METAPHOR
AND SYMBOLIC ACTIVITY

Volume 11, Number 1

10TH ANNIVERSARY SPECIAL ISSUE: FIGURATIVE LANGUAGE AND COGNITIVE SCIENCE

Richard P. Honeck, Guest Editor

EDITOR
Howard R. Pollio, *University of Tennessee*
ASSOCIATE EDITORS
Robert R. Hoffman, *Adelphi University*
Michael K. Smith, *University of Tennessee*
EDITORIAL BOARD
Douglas Berggren, *Yale University*
Richard Billow, *Adelphi University*
Roger Brown, *Harvard University*
Ronald Butters, *Duke University*
Cathy H. Dent-Read, *University of Connecticut*
Arthur Efron, *State University of New York, Buffalo*
Dedre Gentner, *Northwestern University*
Robert Haskell, *University of New England*
Richard Honeck, *University of Cincinnati*
Mark Johnson, *Southern Illinois University*
Rene Jongen, *University of St. Louis, Brussels*
Susan Kemper, *University of Kansas*
John Kennedy, *University of Toronto*
Gary Klein, *Klein Associates, Inc.*
Nathan Kogan, *New School for Social Research*
George Lakoff, *University of California, Berkeley*
Earl Mac Cormac, *North Carolina State University*
David McNeill, *University of Chicago*
David Miall, *College of St. Paul and St. Mary, Cheltenham, England*
Don Nilsen, *Arizona State University*
Andrew Ortony, *Northwestern University*
David Palermo, *Pennsylvania State University*
Robert Romanyshyn, *Pacifica Graduate Institute*
Sylvia Weber Russell, *University of New Hampshire*
Warren Shibles, *University of Wisconsin, Whitewater*
Richard Vondruska, *J. Maloney Associates, Inc.*
Stella Vosniadou, *University of Illinois*
Harald Weinrich, *University of Munich, Germany*
Yorick Wilks, *New Mexico State University*
Ellen Winner, *Boston College*
PRODUCTION EDITOR
Diana Leith, *Lawrence Erlbaum Associates, Inc.*

First published by Lawrence Erlbaum Associates, Inc.

This edition published 2013 by Psychology Press

711 Third Avenue, New York, NY 10017

27 Church Road, Hove East Sussex BN3 2FA

Copyright © 1996, Lawrence Erlbaum Associates, Inc. No part of this publication may be used, in any form or by any means, without permission of the publisher.

Cover: William Blake's *Ancient of Days* is reproduced with the kind permission of the Whitworth Art Gallery, University of Manchester, England.

METAPHOR AND SYMBOLIC ACTIVITY *11*(1), 1–15
Copyright © 1996, Lawrence Erlbaum Associates, Inc.

10TH Anniversary Special Issue: Figurative Language and Cognitive Science

Introduction: Figurative Language and Cognitive Science— Past, Present, and Future

Richard P. Honeck
Department of Psychology
University of Cincinnati

This article, which serves to introduce the companion articles in this issue, briefly reviews the impact of cognitive science on the study of figurative language. The intellectual background for (mainly psychological) inquiry into the topic in the 1970s is described, and a synopsis of various cognitive science contributions up to the present is provided. After focusing on the contemporary view of the impact of cognitive science on figurative language as exemplified by the articles in this issue, a final section looks at some of the problems that lie ahead.

The articles in this special issue focus on experimental psychological, neuropsychological, and computer-based approaches to figurative language. Such variety of disciplines, and of the other disciplines that help inform them, suggests the enormous impact of cognitive science on our understanding of figurative language.

Before addressing the articles in this special issue, I will briefly document some of the changes in the area over the last 25 years, with an eye toward the role of cognitive science. The final section will address some of the significant issues and deficiencies that remain to be resolved, no doubt only through continued interdisciplinary effort.

Requests for reprints should be sent to Richard P. Honeck, Department of Psychology, Room 429, Dyer Hall, University of Cincinnati, Cincinnati, OH 45221–0376. E-mail: richard.honeck@uc.edu.

EXPANSION OF THE FIGURATIVE LANGUAGE AREA: 1970 TO PRESENT

The literature on figurative language has greatly expanded since its first modern and intensive beginnings in the 1970s. Every subarea of cognitive science has contributed. But it was not always this way. A psychologist (like myself) who wanted to study figurative language in the early 1970s could draw on (a) the work of some Gestalt writers, notably Werner (1955) and Werner and Kaplan (1963) on physiognomic perception; (b) Gestalt-influenced studies such as those by Asch (1955) on "double function terms" (e.g., "warm" and "cold") and by Asch and Nerlove (1960) on children's understanding of these terms. A handful of neo-behaviorists examined aspects of figurative language in the 1950s, including Osgood, Suci, and Tannenbaum (1957), who briefly discussed lexical metaphor and synesthesia (e.g., low-pitch tones seem "large," high-pitch tones seem "small"), and Brown (1958), who studied the application of nonauditory words to opera singing.

Aside from this modest inquiry by fellow psychologists, more was to be found in the writings of (a) philosophers, especially Black's (1962) exposition of the interaction view of metaphor, but also newly minted treatments of metaphor within the speech act framework (e.g., Steinmann, 1973); (b) rhetoricians such as Richards (1936), who is responsible for the concepts of *tenor* (topic), *vehicle,* and *ground* as applied to metaphor; (c) a few linguists such as Mathews (1971), who made metaphor a matter of the violation of Chomskyan selection restriction rules, and Chafe (1968), who did the same with idioms. A number of psychologists during this era independently came to the conclusion that figurative language was a strong challenge to existing theories and concepts (see Honeck, 1980, for a fuller exposition of historical matters up to 1980).

A minor explosion of psychological research on figurative language, mainly metaphor, occurred in the 1970s. The impact of experimental psycholinguistics was there, motivated by dissatisfaction with Chomskyan linguistics and a vital interest in problems of meaning representation ("semantic memory"). This tradition is updated in the article by Albert Katz. The Gestalt tradition was less active but was kept viable by some experimentalists and is described in Larry Marks's article. However, neuropsychological studies of figurative language were sparse, a trend that was quickly reversed with a burst of studies in the 1980s and 1990s. Curt Burgess and Christine Chiarello bring us up to date on this fascinating line of inquiry. Prior to 1970, there was practically no interest in figurative language from the artificial intelligence (AI) standpoint. That changed somewhat in the 1970s with the work of Russell (1975) and Wilks (1975) and others, and this upward trend has continued. In this special issue, Jim Martin provides a contemporary overview of this important approach.

By the end of the decade, a number of symposia on metaphor had been held, a seminal book was published by Pollio, Barlow, Fine, and Pollio (1977), a major interdisciplinary conference was convened that resulted in a standard edited volume (i.e., Ortony, 1979b), and an edited volume of largely experimental psychological research appeared (i.e., Honeck & Hoffman, 1980) as did Lakoff and Johnson's (1980) groundbreaking work.

The increasing impact of cognitive science can be found all around us. Texts on figurative language continue to be written by psychologists (e.g., Gibbs, 1994), linguists (e.g., Lakoff, 1987), computer scientists (e.g., Martin, 1992), English studies scholars (e.g., Turner, 1991), and philosophers (e.g., Fogelin, 1988; Mac Cormac, 1985). Texts on the neurology of language have begun to incorporate research findings and concepts about figurative language (e.g., Brownell & Joanette, 1993; Caplan, 1992; Code, 1987). A revision of a standard interdisciplinary volume has been issued (i.e., Ortony, 1993), and an entire edited volume has been devoted to the cognitive science of idioms (i.e., Cacciari & Tabossi, 1993). Last, but not least, this journal was inaugurated in 1986 and remains an important and vital outlet for scholarly works as witnessed by the special issues of 1995 and 1996.

Important theoretical developments have come from every corner of cognitive science. For example, new and competing psychological theories of metaphor comprehension have emerged beyond the somewhat vague, although seminal, entreaties of Black's (1962) interaction view. They include the now older salience imbalance model (Ortony, 1979a) and the domains interaction model (Tourangeau & Sternberg, 1982) but they also include the newer structure mapping model (Gentner, 1983; Gentner & Clements, 1988) and class inclusion model (Glucksberg & Keysar, 1990, 1993). Lakoff and Johnson's (1980) thesis that much of language is built on basic metaphoric–conceptual structures has taken hold and thrives in books (see Johnson, 1987; Lakoff, 1987; Lakoff & Turner, 1989), articles (e.g., Lakoff, 1993), and the new *Cognitive Linguistics* journal, providing a formidable contrast to other approaches. This view has also been adopted as part of one psychology research program (see Gibbs, 1994). Chiarello (1991) offered a neuropsychological "dual semantic systems" model of the roles of the left and right cerebral hemispheres in handling language, including figurative and otherwise "ambiguous" language.

Since the 1970s, metaphor has been the trope of scholarly choice. However, various models and treatments of idioms have appeared (e.g., Cacciari & Tabossi, 1993), all of which indicates that idioms can be syntactically and semantically complex and not mere frozen entries in some mental dictionary. Indeed, there is even recent evidence that totally novel constructions can function idiomatically, even on a first encounter (Forrester, 1995).

Other tropes and tropic functions have received increasing attention. Kumon-Nakamura, Glucksberg, and Brown (1995) reviewed the pragmatic and various echoic theories of irony and then presented evidence for their own allusional pretense theory. Honeck and Temple (1994) presented a revised version of the conceptual base theory of proverbs in the context of a comparison with Lakoff and Turner's (1987) great chain metaphor theory. Honeck and Temple's article was commented on by Gibbs, Colston, and Johnson (in press), with a rejoinder by Honeck and Temple (in press). Gibbs (1995) also wrote about proverbs. Mac Cormac (1985) and Glucksberg, Brown, and McGlone (1993) challenged aspects of the Lakoffian view of metaphor. The research on metonymy, hyperbole, understatement, and simile is scarce, but this seems to be changing (for reviews, see Gibbs, 1993, 1994).

This outpouring of research, of enjoined though internecine battles, the stretching of figurative language boundaries, is all laudable. There are other positive developments: Basic cognitive scientific topics such as analogy (e.g., Barnden & Holyoak, 1994; Gentner & Clements, 1988), categorization (Glucksberg & Keysar, 1993; Honeck, Kibler, & Firment, 1987), perception (Dent, 1984; Kennedy, Green, & Vervaeke, 1993; Paivio & Walsh, 1993; see also the article by Marks in this issue), and the newer personal theories of mind (Winner & Gardner, 1993) have been used to explain aspects of figurative language understanding. New online experimental techniques, that is, those that get closer to real-time processing of language, have been used: priming to study metaphor (e.g., Kemper, 1989) and idioms (e.g., Peterson & Burgess, 1993), positron emission tomography (PET) scans during metaphor processing (Bottini et al., 1994), time to read idiomatic material (e.g., Forrester, 1995), and time to decide whether letter strings are words after first hearing idioms of variable redundancy (e.g., Cacciari & Tabossi, 1988). Articles that attempt to integrate and provide detailed alternative views have appeared (e.g., Glucksberg, 1991). If I have omitted anything in this litany, it is only because of the pure volume of material.

THE PRESENT: DISCUSSION OF THE ARTICLES IN THIS SPECIAL ISSUE

This brings us to the present and to the fine set of essays in this issue. These essays speak for themselves, and I will mention only selected aspects of their content.

Katz frames his discussion in terms of developments in experimental psycholinguistics since the time of the Ortony, Reynolds, and Arter (1978) review article. At that time, psycholinguists had already begun to consider the role of pragmatic factors in figurative understanding, most especially context, but also in speech act theory and the Gricean cooperative principle. Speaker intentions, context, background knowledge, and common cognitive ground were also being considered. And, if anything, this loose "everything counts model" motivated the ensuing 20-year psycholinguistic search for the Holy Grail—how do people understand figurative language if the understanding is not to be found in the literal meaning alone?

Of course, psychological research had already established that everything counts. The legacy of the Chomsky era of the 1960s was that people construct complex wholistic structures beyond deep structure using language as a complex guide for various depths and kinds of inferences. In the 1970s, figurative language scholars embraced the everything counts model in conjunction with the standard pragmatic model, whereas in the 1980s, a more radical contextualism took hold, only to give rise, I believe, to a more moderate literalism in the 1990s. In any event, the initial experimental work on figurative language was already constrained to look beyond the literal sentence level for answers. Despite this and the ensuing historical changes, the leitmotif in psychological studies of figurative language has revolved around literalness.

It is no accident, therefore, that essentially all of the issues raised by Katz stem from this thematic core. Even the a priori question of why tropes are used can be construed as a question of why literal utterances are not enough. The mental representation issue is divided by Katz into questions of whether (the literal) imagery evoked by utterances plays a central role in figurative processing, theories of what kinds of mental structures get mutually mapped as part of metaphor processing, and how much the (literal) parts of a trope contribute to its figurative meaning. The final issue raised by Katz—that of exactly what information is processed—also focuses on the relative contribution of literal information to trope recognition and understanding. Actually, Katz first reviews some of the experimental techniques used to elucidate figurative processing. Again, it is notable that many of the newer online techniques that he mentions—probe reaction time, priming, and so forth—were developed to get a closer look at literal compositional processes.

What is the organismic source of figurative language? Because nonlinguistic stimuli can evoke figurative reactions, there is some reason for seeking the answer in basic psychological processes, especially perception and perhaps biology. Marks pursues one line of research in this realm, namely that on synesthesia and cross-modal similarities of various kinds. Although the hypothesis of innate figurative perception was promulgated by Gestalt psychologists, Marks extends this tradition via a more experimentalist perspective.

After distinguishing between "strong synesthesia," which is rare, and "weak synesthesia," which is much more common and presumably manifest in making nonliteral cross-modal connections, Marks cites evidence for the reliability of certain of these connections (e.g., brightness and pitch), a reliability that he asserts is hard wired. The remainder of the article focuses on the question of whether all sensory stimuli evoke a generalized physiognomic perceptual response that mediates synesthetic and cross-modal equivalences. Reliable physiognomic connections between stimuli abound—strongness with darker colors, happy emotions with lighter colors, visual shapes with particular names (e.g., the famous "maluma" and "takete"), speech sounds with particular meanings, words (e.g., sneeze vs. cough) with visual brightness, and so on. Newer research reported by Marks indicates that physiognomic effects occur quite rapidly and that the identification of a stimulus (e.g., a U-shaped stimulus) can be speeded up by the copresentation of a synesthetically related stimulus (e.g., a low-pitched tone). Studies of this sort fit into the current emphasis in the literature on online measures, relatively automatic processing, and the preattentive and noneffortful effects of priming or priming-like mechanisms.

A major theoretical issue taken up by Marks is whether linguistic processes play a role in cases of cross-modal connections. He frames the issue in terms of the semantic hypothesis versus the physiognomic hypothesis. The former holds that there is a common, linguistically mediated basis for all cross-modal interactions; the latter holds that there is an "early, undifferentiated" (p. 55) perceptual response of a physiognomic sort, linguistic processes being neither necessary nor sufficient for the response to occur. Marks reports results that provide some support for the semantic hypothesis. However, because figurative perception can occur in preverbal children, and quite rapidly in any case, Marks speculates that all "expressive meanings" may be handled by a "single, common amodal mechanism" (p. 59). I might add that, whether it is a mechanism or a general organismic proclivity, there is in the feeling tone quality of physiognomic perception a powerful basis for trope production and comprehension.

From a "nonwet" biological view we move to concrete brain realities. What can be said of the linguistic skills of the two cerebral hemispheres? In general, the left hemisphere is better prepared to produce and comprehend phonological and syntactic information and, perhaps, semantic information as well. The left brain also predominates in praxis, or sequential, goal-oriented bodily movements that are not under spatial environmental constraints (Corballis, 1991). Corballis asserts that the left brain has a "generative assembling device" (GAD) that is capable of synthesizing new forms and meanings by using a small number of elements, whether linguistic, musical, visuo-graphic, abstract ratiocinative, or whatever. This is a big theoretical step however, and not everyone would agree with it.

The right hemisphere seems to have an advantage in certain nonlinguistic skills such as recognizing familiar objects and faces, depth perception and other basic perceptual skills, nonspeech auditory inputs, mental rotation of figures, spatial attention, and certain aspects of emotion processing and production including prosodic and feeling tone concomitants of speech (e.g., see Corballis, 1991, for a review). As Corballis put it, "many of the functions that seem to be associated more with the right hemisphere . . . are precisely the sorts of functions for which words often fail us" (p. 274). I find this statement interesting not simply because it is insightful but also because of what it does not say—namely that Corballis does not consider right-brain involvement in the comprehension of language that can be understood in more than one way, on different levels, or for which, in general, the entire informational envelope for a linguistic utterance must be considered. Note, however, that right-brain involvement in language is a controversial issue.

This brings us to the specifics of the Burgess and Chiarello article. These authors first describe some evidence that the right hemisphere not only has good comprehension abilities but that nonobvious, subordinate, or "weak" meanings are more likely to be aroused in the right brain and that they tend to stay activated longer. The framework for their discussion is that both bottom-up (largely semantic) and top-down (largely pragmatic) factors must be jointly used to explain figurative processing by both brains. That the right brain is heavily implicated in metaphor comprehension is indicated by a number of results. Patients with right-hemisphere damage (RHD) are more likely than patients with left-hemisphere damage (LHD) to match metaphors with a depiction of their literal meaning rather than their figurative meaning, provide literal rather than figurative interpretations of metaphors, and match double function terms such as *warm* and *deep* with other words by using denotative rather than connotative or metaphorical criteria. There is even PET scan evidence for greater right than left brain activity while processing sentence metaphors.

A similar, although perhaps more intriguing, picture emerges for idiom processing. Here, Burgess and Chiarello suggest that because the left brain is more likely than the right to get a quick interpretation of an input and stick with it, idioms may cause it problems because it is, " 'garden-pathed' away from a figurative interpretation during processing" (p. 74). The initial words of an idiom (e.g., "kick the") may therefore trigger conventional or literal processing that is not revised when later words (e.g., "bucket") are encountered. However, this view cannot explain why the left brain could not have stored the entire idiom, qua idiom, in the first place. Burgess and Chiarello aver that processing differences between the two brains, not representational differences, should be emphasized and hypothesize that because the right brain does a thematic rather than a syntactic analysis of linguistic inputs, it is better able to access the idiom and its associated meaning in a wholistic way by a "bootstrap" operation. Last, the authors cite evidence that the two brains may be equally capable at the earliest stages of language processing as indicated by no differences among patients with RHD, patients with LHD, and normal controls on an online idiom processing task, whereas normal participants performed better than the other two groups on an idiom interpretation task. This complication suggests that the more typical inferior performance of patients with RHD reflects a later stage of language processing.

Burgess and Chiarello conclude with thoughts on pragmatics and the ever-present issue of intention. They begin by noting that cerebral asymmetries are more likely in word recognition and priming studies when late emerging, controlled attentional processes are recruited. Bottom-up semantic processes occur earlier, however, which leads to the conclusion that figurative understanding may have more to do with later, pragmatic processes including intention. They allude here to a study that found that patients with RHD had trouble distinguishing outright lies from ironic jokes. A liar intends to be believed, whereas the ironic joke teller does not. Such a result is consistent with the growing literature on the deficiencies of patients with RHD in processing humor, indirect speech acts, proverbs, and a large assortment of indirect forms of language in general (Brownell & Joanette, 1993; Chiarello, 1991; Code, 1987; see also the article by Burgess and Chiarello in this issue).

The Burgess and Chiarello article focuses on experimental studies of brain damaged individuals. There are tantalizingly related case studies of individuals who suffer from apparent but unspecifiable brain damage. I can only briefly allude to some interesting examples. First, individuals with autism are notoriously literal in their use of language. It is not clear why, but one explanation is that they lack a theory of mind, a conception, that is, of other people's beliefs, intentions, and kinds of understandings (Baron-Cohen, Leslie, & Frith, 1985). Such a deficiency might well prevent the autistic person from understanding and using figurative language because the hallmark of figurativeness is that something is said that is not meant conventionally and that may violate conversational norms. If one cannot understand that someone else could intend one meaning but say something else, then one might perform much as the patients with RHD described by Burgess and Chiarello.

An interesting complement to these musings is the case described by Sacks (1995) of "TG," a college professor with Asperger's syndrome, a "higher functioning" form of autism. Sacks suggests that TG operates largely on a literal and other-mind-less level—her writing has gaps and discontinuities that suggest that she fails to keep in mind her reader's minds, and she has problems deciphering irony, metaphor, proverbs, and other forms of nonliteral usage.

From the developmental neuropsychological literature come equally fascinating cases of the double dissociation of language and cognition. These include "chatterboxes" who are quite grammatical and adept in their loquaciousness but otherwise retarded (see Pinker, 1994) or children with Williams syndrome, a metabolic disorder that leads, among other things, to good grammatical abilities, the use of unusual words, and fair semantic–lexical knowledge, coupled with intellectual retardation (Bellugi, Bihrle, Neville,

Doherty, & Jernigan, 1992). Studies using event-related potentials with participants with Williams syndrome indicate abnormal responses, including an abnormally large response to semantic anomaly in priming studies (Bellugi et al., 1992) about which the researchers state that "the effect of greater priming may indicate an abnormally large degree of activation which extends to less frequent and nonprototypical items within the mental lexicon" (p. 226). No data are reported on these participants' skill in handling figurative language, but if we accept the idea that the right hemisphere activates less dominant meanings and that these meanings linger and that this facilitates figurative processing, then one might expect fair comprehension in this realm.

Finally, there is the recent case study reported by Smith and Tsimpli (1995) of Christopher, now in his early thirties. Christopher has command of not just English, but many other languages and can translate between many of them but tends to score in the retarded range on tests of nonverbal intelligence. What is significant for our purposes is that he does not have what the authors call "second order" or "metarepresentation" skills, including the ability to decode jokes, irony, or metaphor, to appreciate the difference between use and mention, and to understand such things as plans for a future state of affairs as in, "Pete said that Joan is going to Chicago tomorrow." It is curious as well that Christopher seems constitutionally unable to tell a lie.

Like the other cases just cited, this uneven profile of language and intellect raises provocative questions about the relation between cerebral asymmetries, modularity, intelligence, and figurative processing.

From brains and language we move to AI, the attempt to get computers to do what comes naturally for brains. It is fair to say that of all the cognitive science disciplines and when considering figurative language, AI has had the least impact on the others. AI has been an "outlier." The reasons for this are many and complex. I list here some reasons that together provide a stereotype of AI work by non-AI scholars. By this stereotype, AI scholars (a) receive different training than do scholars coming from the arts and sciences; (b) often use idiosyncratic concepts; (c) seem more concerned with the constraints imposed by the technicalities of computers than with the realities of the human system such as attention and working memory; (d) often seem unconcerned with empirical constraints or even the real world and how it relates to symbols that, ironically, are manipulated by elegant formalisms; (e) use programming language, jargon, special and arcane symbols, and conventions in general that are interpretable in some cases only after intensive study; (f) finesse the crucial issues such as the psychologically real format of "data"; (g) seem to approach problems with a different metatheoretical stance, including the fantastic idea that something as human as metaphor production and comprehension could be carried out by a computer; and (h) are even ordinarily housed somewhere far away from other cognitive scientists on a campus. In short, and so the stereotype goes, practitioners of AI are in their own world, use their own concepts, ignore important issues, and in general do things their own way someplace where the rest of us cannot find them and vice versa.

There is both truth and fiction in the foregoing stereotype. However, it may be more appealing to non-AI scholars of figurative language than to cognitive scientists who work in other areas. That is why I believe that the article by Martin should be helpful to those in the former category.

Martin begins by telling us that figurative language causes explanatory problems because it violates the compositionality principle, and even when all sorts of top-down information about meaning, context, and so forth are factored in, figurative language comes out of computer programs looking like "ill-formed," unacceptable-looking "weird stuff." To deal with this situation, three distinct approaches to figurative language have evolved.

The *convention-based* approach basically associates one kind of fact, concept, or mental representation with another. Martin illustrates this approach, for example, with metaphor, for which the program uses conceptual metaphors, perhaps Lakoffian ones. Various processes go into action once the source of the conceptual metaphor is matched, and a different conception of the surface level target is created. For idioms, this approach may work by simply associating the idiom with a particular meaning, whether derived semicompositionally or not. For indirect speech acts, the literal meaning might be associated with the conveyed meaning by a general rule.

The second approach is *reasoning based.* In this approach, compositionality is given some credence, and general cognitive mechanisms, such as analogy, are used to solve the figurative puzzle. For metaphor, analogies may be used; for indirect speech acts, questions may be asked about information shared by the speaker and listener. For metonymy (e.g., "The lettuce-bacon-tomato forgot to pay his bill"), spreading activation mechanisms that come to the "conclusion" that "lettuce-bacon-tomato" refers to a person who forgot to pay the bill may be put to work.

Third, there is the *context-based* approach. In this approach, prior discourse is given a predominant role in informing the current context, which, in turn, generates likely meanings. The figurative utterance itself is seen only as "a cue to be used to select and possibly flesh out some part of a current discourse representation" (p. 86). This approach makes metaphor nonspecial—just more information that can be used to inform a developing mental model for the discourse. In some varieties of this approach, metaphors are related to memories of specific prior contexts.

Martin notes that computer scientists have taken different tacks within each of the three approaches and that they have been variously successful in handling a particular kind of trope. Also, a theme that pops up throughout his discussion is that tropes of every kind, idioms included, exhibit great fluidity, productivity, and context sensitivity. He concludes by lamenting the absence of a general integrative framework and suggests that scholars in the area take empirical constraints more seriously. Surely, Martin's discussion should help dispel the hyperbole of the previously described stereotype.

THE FUTURE: LINGERING ISSUES AND DEFICIENCIES

Although the cognitive science nexus has spurred progress on figurative language, old issues remain and gaps of various kinds have emerged. Several issues can be briefly noted as follows:

1. *Roots.* Where does figurative language, whether inborn perceptual proclivities, Lakoffian conceptual metaphoric structures, or the entire processing system come from? Or, is the question badly put, with the organism innately driven to express one thing in terms of another, using whatever learned means are available?
2. *Literalness.* How are literal, figurative, conventional, and related concepts alike and different? Much has been written about this, but are we any closer to a consensual resolution?
3. *The standard pragmatic model.* Does the language receiver go through a series of stages from literal to figurative in accordance with the stipulations of speech act theory, the cooperativeness principle, and the like? Is the model more likely to work in some situations for some tropes than for others?
4. *Similarity.* We all use this concept when discussing simile, metaphor, proverbs, and so forth, but it seems to be an effect rather than a cause. What theoretical status should it be given?
5. *Context.* Like the weather and similarity, we all talk about it, but there is no overarching model of how it does all the things that are attributed to it. But we need it or something like it.
6. *Metacognition.* This is the issue of what people know of their own and other people's minds, especially the latter when it comes to figurative language. What role does it play for different tropes, for different purposes, and on different occasions?
7. *Processing mode.* Is there a special figurative processing mode, a mental set to take things either literally or nonliterally? It is clear that experimental participants develop processing modes and, therefore, experimentalists have resorted to various strategies to handle this but they are probably inadequate because participants are often very good at divining, especially, the figurative qualities of verbal inputs designed by experimentalists to be processed literally.

Processing mode and context are more changeable, temporary conditions of the person and the person's circumstances, respectively. Roots and metacognition are less changeable, more clearly long-term learning factors. Literalness, stages, and similarity all seem to result from the interaction of the other factors. And that is the rub—potentially all of these factors interact! Context is well-known for making utterances look literal or nonliteral, but it is not necessary. Frame of mind can do this on its own. Without certain roots, however, no amount of context may be helpful, even when metacognition is advanced. If it is not advanced, then figurative processing may be impossible, in which case the construction of certain abstract kinds of similarity, and the stages of processing, become limited. Not knowing how these factors are operative for a given individual makes it practically impossible to make judgments about whether something was literal or not, conventional or not, figurative or not, similar or not. Context, roots, metacognition, and familiarity with certain kinds of tropes may make processing look so automatic and stage free that no currently existing measuring techniques could elucidate the stages, if there were any. If we cannot say what is literal with any certainty, then how do we assess models that either assume or do not assume its priority? And so on. Clearly, it is almost impossible to discuss any one of these several issues without simultaneously bringing more of them into the picture. The picture is sobering, but it largely defines the theoretical task ahead—everything does count!

Part of the puzzle of this picture is due to gaps in our implementation of research on figurative language. It is generally recognized that there have been limitations on the kind (mainly metaphor and idioms), complexity (not very), and linguistic format (e.g., "isa" metaphors) of the tropes examined.

Furthermore, there is, I believe, some confusion about tropic form and function. The basic forms I take to be idiom, metonymy, simile, metaphor, oxymoron, and proverb. Each one of these forms can perform a large variety of functions, including irony, sarcasm, indirect speech act, hyperbole, understatement, rhetorical question, and more. Theoretically, we could construct a table with forms on the rows and functions on the columns, and the table could be filled with examples. So on one level, form and function are orthogonal, but in terms of actual usage, they are not. Certain tropes are more likely to be used for certain purposes, such as using metaphor to convey irony and proverbs as understatement (see Roberts & Kreuz, 1994, for an empirical study of why people use various forms of figuration).

Finally, I suspect that the feeling of uneasiness some scholars have about the area can be traced not just to the lack of integrative theory—a commonplace—but to the lack of a larger framework. Where, exactly, does figurative language fit in the larger scheme of things? My favorite idea here is "indirectness," because this is a quality that all tropes have but also because it brings in other kinds of cognition, both linguistic and nonlingistic. In this regard, I have used figurative language as a springboard for discussing intelligence and mental retardation (Honeck, in press). The basic thesis is that much of intelligent behavior partakes of indirectness—layeredness, levels, embeddedness, and tangled heirarchies. Deficiencies in this respect brought about by inabilities to build processing structures by using prior or old information, result in a lesser ability to process not only figurative language but all sorts of indirect cognition including pretense, deception, and so forth. In any event, we need to "un-isolate" figurative language and make it less special, whether by viewing it in the more general light of indirectness or some other rubric. Perhaps this special issue can further this goal.

REFERENCES

Asch, S. (1955). On the use of metaphor in the description of persons. In H. Werner (Ed.), *On expressive language* (pp. 29–38). Worcester, MA: Clark University Press.

Asch, S., & Nerlove, H. (1960). The development of double function terms in children. In B. Kaplan & S. Wapner (Eds.), *Perspectives in psychological theory* (pp. 47–60). New York: International Universities Press.

Barnden, J., & Holyoak, K. (Eds.). (1994). *Analogy, metaphor, and reminding.* Norwood, NJ: Ablex.

Baron-Cohen, S., Leslie, A. M., & Frith, U. (1985). Does the autistic child have a 'theory of mind'? *Cognition, 21,* 37–46.

Bellugi, U., Bihrle, A., Neville, H., Doherty, S., & Jernigan, T. (1992). Language, cognition, and brain organization in a neurodevelopmental disorder. In M. R. Gunnar & C. A. Nelson (Eds.), *The Minnesota symposia on child psychology: Vol. 24. Developmental behavioral neuroscience* (pp. 201–232). Hillsdale, NJ: Lawrence Erlbaum Associates, Inc.

Black, M. (1962). *Models and metaphors.* Ithaca, NY: Cornell University Press.

Bottini, G., Corcoran, R., Sterzi, R., Paulesu, E., Schenone, P., Scarpa, P., Frackowiak, R. S. J., & Frith, C. D. (1994). The role of the right hemisphere in the interpretation of figurative aspects of language: A positron emission tomography activation study. *Brain, 117,* 1241–1253.

Brown, R. (1958). *Words and things.* Glencoe, IL: Free Press.

Brownell, H. H., & Joanette, Y. (Eds.). (1993). *Narrative discourse in neurologically impaired and normal aging adults.* San Diego: Singular.

Cacciari, C., & Tabossi, P. (Eds.). (1993). *Idioms: Processing, structure, and interpretation.* Hillsdale, NJ: Lawrence Erlbaum Associates, Inc.

Caplan, D. (1992). *Language: Structure, processing, and disorders.* Cambridge, MA: MIT Press.

Chafe, W. (1968). Idiomaticity as an anomaly in the Chomskyan paradigm. *Foundations of Language, 4,* 109–127.

Chiarello, C. (1991). Interpretation of word meanings by the cerebral hemispheres: One is not enough. In P. J. Schwanenflugel (Ed.), *The psychology of word meanings* (pp. 251–278). Hillsdale, NJ: Lawrence Erlbaum Associates, Inc.

Code, C. (1987). *Language, aphasia, and the right hemisphere.* New York: Wiley.

Corballis, M. C. (1991). *The lopsided ape: Evolution of the generative mind.* New York: Oxford University Press.

Dent, C. (1984). The developmental importance of motion information in perceiving and describing metaphoric similarity. *Child Development, 55,* 1607–1613.

Fogelin, R. J. (1988). *Figuratively speaking.* New Haven, CT: Yale University Press.

Forrester, M. A. (1995). Tropic implicature and context in the comprehension of idiomatic phrases. *Journal of Psycholinguistic Research, 24,* 1–21.

Gentner, D. (1983). Structure mapping: A theoretical framework for analogy. *Cognitive Science, 7,* 155–170.

Gentner, D., & Clements, C. (1988). Evidence for relational selectivity in the interpretation of analogy and metaphor. In G. Bower (Ed.), *The psychology of learning and motivation* (Vol. 22, pp. 307–358). Orlando, FL: Academic.

Gibbs, R. W. (1993). Process and products in making sense of tropes. In A. Ortony (Ed.), *Metaphor and thought* (2nd ed., pp. 252–276). New York: Cambridge University Press.

Gibbs, R. W. (1994). *The poetics of mind: Figurative thought, language and understanding.* New York: Cambridge University Press.

Gibbs, R. W. (1995). What proverb understanding reveals about how people think. *Psychological Bulletin, 118,* 133–154.

Gibbs, R. W., Colston, H., & Johnson, M. (in press). Proverbs and the metaphorical mind. *Metaphor and Symbolic Activity.*

Glucksberg, S. (1991). Beyond literal meanings: The psychology of allusion. *Psychological Science, 2,* 146–152.

Glucksberg, S., Brown, M., & McGlone, M. (1993). Conceptual metaphors are not automatically accessed during idiom comprehension. *Memory & Cognition, 21,* 711–719.

Glucksberg, S., & Keysar, B. (1990). Understanding metaphorical comparisons: Beyond similarity. *Psychological Review, 97,* 3–18.

Glucksberg, S., & Keysar, B. (1993). How metaphors work. In A. Ortony (Ed.), *Metaphor and thought* (2nd ed., pp. 401–424). New York: Cambridge University Press.

Honeck, R. P. (1980). Historical notes on figurative language. In R. P. Honeck & R. R. Hoffman (Eds.), *Cognition and figurative language* (pp. 25–42). Hillsdale, NJ: Lawrence Erlbaum Associates, Inc.

Honeck, R. P. (in press). An outsider looks at mental retardation: A moral, a model, and a metaprinciple. In N. Bray (Ed.), *International review of research in mental retardation.* New York: Academic.

Honeck, R. P., & Hoffman, R. R. (Eds.). (1980). *Cognition and figurative language.* Hillsdale, NJ: Lawrence Erlbaum Associates, Inc.

Honeck, R. P., Kibler, C., & Firment, M. J. (1987). Figurative language and psychological views of categorization: Two ships in the night? In R. E. Haskell (Ed.), *Cognition and symbolic structures: The psychology of metaphoric transformation* (pp. 103–120). Norwood, NJ: Ablex.

Honeck, R. P., & Temple, J. G. (1994). Proverbs: The extended conceptual base and great chain metaphor theories. *Metaphor and Symbolic Activity, 9,* 85–112.

Honeck, R. P., & Temple, J. G. (in press). Proverbs and the complete mind. *Metaphor and Symbolic Activity.*

Johnson, M. (1987). *The body in the mind: The bodily basis of meaning, reason and imagination.* Chicago: University of Chicago Press.

Kemper, S. (1989). Priming the comprehension of metaphors. *Metaphor and Symbolic Activity, 4,* 1–17.

Kennedy, J. M., Green, C. D., & Vervaeke, J. (1993). Metaphoric thought and devices in pictures. *Metaphor and Symbolic Activity, 8,* 243–255.

Kumon-Nakamara, S., Glucksberg, S., & Brown, M. (1995). How about another piece of pie: The allusional pretense theory of discourse irony. *Journal of Experimental Psychology: General, 124,* 3–21.

Lakoff, G. (1987). *Women, fire and dangerous things: What categories reveal about the mind.* Chicago: University of Chicago Press.

Lakoff, G. (1993). The contemporary theory of metaphor. In A. Ortony (Ed.), *Metaphor and thought* (2nd. ed., pp. 202–251). New York: Cambridge University Press.

Lakoff, G., & Johnson, M. (1980). *Metaphors we live by.* Chicago: University of Chicago Press.

Lakoff, G., and Turner, M. (1989). *More than cool reason: A field guide to poetic metaphor.* Chicago: University of Chicago Press.

Mac Cormac, E. R. (1985). *A cognitive theory of metaphor.* Cambridge, MA: MIT Press.

Martin, J. H. (1992). *A computational model of metaphor interpretation.* Cambridge, MA: Academic.

Mathews, R. (1971). Concerning a linguistic theory of metaphor. *Foundations of Language, 7,* 413–425.

Ortony, A. (1979a). Beyond literal similarity. *Psychological Review, 86,* 161–180.

Ortony, A. (Ed.). (1979b). *Metaphor and thought.* New York: Cambridge University Press.

Ortony, A. (Ed.). (1993). *Metaphor and thought* (2nd ed). Cambridge, England: Cambridge University Press.

Ortony, A., Reynolds, R. E., & Arter, J. A. (1978). Metaphor: Theoretical and empirical research. *Psychological Bulletin, 85,* 919–943.

Osgood, C. E., Suci, G. J., & Tannenbaum, P. H. (1957). *The measurement of meaning.* Urbana: University of Illinois Press.

Paivio, A., & Walsh, M. (1993). Psychological processes in metaphor comprehension and memory. In A. Ortony (Ed.), *Metaphor and thought* (2nd. ed., pp. 307–328). New York: Cambridge University Press.

Peterson, R. R., & Burgess, C. (1993). Syntactic and semantic processing during idiom comprehension: Neurolinguistic and psycholinguistic dissociations. In C. Cacciari & P. Tabossi (Eds.), *Idioms: Processing, structure, and interpretation* (pp. 201–225). Hillsdale, NJ: Lawrence Erlbaum Associates, Inc.

Pinker, S. (1994). *The language instinct: How the mind creates language.* New York: Morrow.

Pollio, H., Barlow, J., Fine, H., & Pollio, M. (1977). *Psychology and the poetics of growth: Figurative language in psychology, psychotherapy, and education.* Hillsdale, NJ: Lawrence Erlbaum Associates, Inc.

Richards, I. A. (1936). *The philosophy of rhetoric.* New York: Oxford University Press.

Roberts, R. M., & Kreuz, R. J. (1994). Why do people use figurative language? *Psychological Science, 5,* 159–163.

Russell, S. W. (1975). Computer understanding of conceptually complex phrases. *Dissertation Abstracts International, 36,* 2355B–2356B. (University Microfilms No. 75–25, 600).

Sacks, O. (1995). *An anthropologist on Mars: Seven paradoxical tales.* New York: Knopf.

Smith, N., & Tsimpli, I. (1995). *The mind of a savant: Language learning and modularity.* Cambridge, MA: Blackwell.

Steinmann, M. (1973). Figurative language and the two-code hypothesis. In R. W. Fasold & R. W. Shuy (Eds.), *Analyzing variation in language* (pp. 220–227). Washington, DC: Georgetown University Press.

Tourangeau, R., & Sternberg, R. (1982). Understanding and appreciating metaphors. *Cognition, 11,* 203–244.

Turner, M. (1991). *Reading minds: The study of English in the age of cognitive science.* Princeton: Princeton University Press.

Werner, H. (1955). A psychological analysis of expressive language. In H. Werner (Ed.), *On expressive language.* Worcester, MA: Clark University Press.

Werner, H., & Kaplan, B. (1963). *Symbol formation.* New York: Wiley.

Wilks, Y. (1975). Preference semantics. In E. Keenan (Ed.), *The formal semantics of natural language.* New York: Cambridge University Press.

Winner, E., & Gardner, H. (1993). Metaphor and irony. In A. Ortony (Ed.), *Metaphor and thought* (2nd ed., pp. 425–446). New York: Cambridge University Press.

[illegible]

METAPHOR AND SYMBOLIC ACTIVITY *11*(1), 17–37
Copyright © 1996, Lawrence Erlbaum Associates, Inc.

Experimental Psycholinguistics and Figurative Language: Circa 1995

Albert N. Katz
Department of Psychology
The University of Western Ontario

This article is a review of the methodology employed and of selected research questions found in the current experimental psycholinguistic literature on figurative language. A contrast is made between methodologies based on memory for language and those based on moment-by-moment (online) measures of processing. Research questions examined include (a) why people might use figurative language instead of a literal translation, (b) the nature of the mental representation assumed to underlie figurative language, (c) how figurative meaning is computed, and (d) the role played by conceptual structures in the understanding of tropes.

Ortony, Reynolds, and Arter (1978) wrote what was at the time the definitive statement on the status of metaphor theory and research in psychology. Looking back at that paper with the hindsight of a later generation, one is struck by how few empirical studies addressed the issue of adult figurative language in those days, the methodological limitations of the studies that had been performed, and the general failure to integrate the research into the larger psycholinguistic literature although attempts were being made by some to relate their work to then-current models of semantic memory. The 17 years since the publication of Ortony, Reynolds, and Arter has seen an explosion in the number of papers published and sophistication in the methodology with which issues in figurative language have been addressed. One aim of this article is to review the methodology currently employed in the study of figurative language.

One is also struck by the prescience of Ortony, Reynolds, and Arter (1978). Metaphor, they argued, has to be seen as a context-bound phenome-

Requests for reprints should be sent to Albert N. Katz, Department of Psychology, The University of Western Ontario, London, Ontario, Canada, N6A 5C2. E-mail: katz@sscl.uwo.ca.

non and by implication described, in part, at a pragmatic level. The accepted opinion of the day—that literal language had processing priority—was seen as problematic. Moreover, they recognized the need for a process explanation for the transition from the expressed literal sense of an utterance to its intended, figurative meaning. A second aim of this article is to describe some of the attempts that have been made over the last 17 years to address these and other questions in the adult psycholinguistic literature.

METHODOLOGICAL ISSUES IN THE STUDY OF FIGURATIVE LANGUAGE

Level of Analysis

In the broadest sense, an utterance can be understood as figurative when the expressed meaning differs from the meaning one intends to convey. This can occur at different levels of analysis: at the level of an individual word, the sentence, the discourse context in which the sentence is presented, and the larger, historical–cultural bases for understanding the trope.

Consider, for instance, an idiom such as *"spilling the beans."* Our understanding of the phrase has to be based, in part, on our understanding of the individual words that constitute the idiom, as suggested by the fact that idiomatic sense is altered when the original terms are replaced with words of similar meaning (e.g., *"scatter the beans"; "spill the peas"*). An ongoing debate is the role, the conditions, and even the necessity, for elaboratively processing each word in a figurative utterance.

Figurative statements usually are not localized just at the word level and, in fact, most theorists have placed the conceptual load at the sentence level. A clear statement of the general problem was expressed by Searle (1979). When a person asserts *"S is a P"* (e.g., *"Sam is a pig"*), intending to convey that *S is an R* (e.g., *Sam is filthy*), one not only has to recognize that the expressed assertion is not intended, but that one must also compute the intended meaning from characteristics of the expressed assertion itself. This approach, commonly described as the standard pragmatic model, asserts that literal meaning *(S is P)* must be processed and that figurative meaning *("S is R")* is optional and will only be initiated when the literal sense is found to be defective. Much of the research over the past 17 years has revolved around questions arising from this perspective.

Ortony, Reynolds, and Arter (1978) were among the first to empirically address effects that could not be localized at the level of the individual sentence. They described the problem of interpreting sentences such as "Regardless of the danger, the troops marched on." Given Searle's (1979) analysis, the conflict between intended and expressed meaning is not found in semantic properties of the words being related in a sentence, but rather between the sentence and the discourse context in which it is embedded. In one case, *S is a P* is an adequate resolution of information derived from the discourse (such as when *the troops* refers to soldiers in an army), whereas, in another context, the literal resolution is inadequate, such as when *"the troops"* refers to unruly children (and the interpretive problem becomes one of identifying an adequate *"S is R"* resolution). Ortony, Schallert, Reynolds, and Antos (1978) demonstrated that placing sentences such as that describing "troops," in a sufficiently long discourse context eliminated the difference in the time taken to read the sentence, whether presented in the *"S is a P"* or *"S is an R"* form. At one level, these data have been taken as evidence against Searle's proposition that literal meaning must be processed before figurative meaning is computed. In a more general sense, the use of more elaborated context has led to questions about the various roles that context plays in figurative language comprehension.

Finally, there is an even more abstract level of analysis: the (often tacit) conceptual underpinnings of meaning. For instance, Lakoff and Johnson (1980) argued that natural language presumes and expresses conceptual meanings represented in some basic root metaphors. As an example, consider the following statements:

"He swallowed the whole argument."
"It took her a long time to digest the new concept."
"They regurgitated the professor's pet theory."

These statements, although quite different, can be tied together by a common underlying, or root metaphor, namely, *"IDEAS ARE FOOD."* Some have argued that the ability to comprehend statements involves making contact with these conceptual metaphoric structures, and the ability to comprehend or generate new metaphors can be understood by an extension of the root metaphor. In Lakoff and Johnson, language is dominated by metaphor and not, as in the standard pragmatic theory, by literal language. Over the last 17 years there have been several attempts to more clearly specify the conceptual domains relevant to explaining how metaphor and other tropes are produced and understood.

Experimental Approaches to the Study of Metaphor

In cognitive science, one can distinguish between studies that attempt to determine functional constraints from those studies that attempt to confirm (or disconfirm) a specific processing algorithm. Presumably, any psychologically realistic theory of metaphor would have to account for these constraints. However, because different algorithms might be able to provide a coherent account of the constraints that are operating, the task for an experimentalist is to determine which of the algorithms are most psychologically real or the conditions under which the algorithm is engaged. For instance, it has long been known that reversing the topic and vehicle of a metaphor can have extremely large effects on the meaning (and sometimes even the grammaticalness) of the sentence (e.g., Ortony, 1979).

Consider, for instance, the shifts in meaning that occur in the following two metaphors:

> *"My surgeon is a butcher."*
> *"My butcher is a surgeon."*

Without discussing the effect in detail, one can note that any psychologically real theory is constrained to account for the shifts in meaning that occur with reversal of topic and vehicle. In principle, several different algorithms, or process models, could account for the same set of constraints. In fact, Ortony (1979) and Glucksberg and Keysar (1990) differ even in the constraints that such a process theory would have to meet.

The methodology employed in a study is, by necessity, limited to the level of analysis of interest and whether the theoretical interest is in isolating functional constraints or testing alternative algorithms. Two basic classes of methods are in popular use: memory-based methods and online methods.

Memory-Based Techniques

There are two general variants based on memory. In one variant, the information of importance is stored as permanent knowledge (of semantics, of beliefs about reality, etc.), whereas in a second variant, the information of importance involves memory for an episode, such as that which occurs when one is asked to remember a previously read story. *Judgment tasks* exemplify the former permanent memory approach. In a typical study, participants are given word strings and asked to make some judgment on the string such as its grammaticality, aptness as a metaphor, centrality to the discourse in which it is embedded, and so forth. The judgments are then analyzed for the linguistic knowledge that the participant brings to the language tasks. For instance, norming studies have shown that participants reliably rate word strings along dimensions such as metaphoricity, familiarity, imagery level, and aptness (e.g., Katz, Paivio, Marschark, & Clark, 1988). Presumably these ratings reflect stored knowledge about figurative language and its uses.

In some studies, judgment data are employed as a means of choosing items to use as independent variables in experiments. As one example, participants might be asked to process metaphors that differ in familiarity, metaphors in which familiarity is based on previously normed word strings. In other experiments, these judgments are used as dependent variables to measure the effect of a manipulation. In a typical study of this type, participants are presented stimuli that are constructed to test some theoretical claim, and the experimenter then examines whether changing the structure of the sentence leads to theoretically expected changes in intuition such as shifts in metaphor aptness or comprehensibility (e.g., Katz, 1982; Trick & Katz, 1986).

Episodic memory testing has been employed as a second form of memory-based methodology. In the typical experiment, participants are presented some target information (e.g., a sentence either out of context or within some discourse) and, at a later time, are tested on their memory for the target. The relation between comprehension and memory is quite high, and consequently, the memory scores are often taken as an indirect index of the processes that went into comprehension. For instance, if memory for a literal sentence (e.g., *You can make an opportunity available to a person, but you can't force him to take advantage of it.*) is contrasted with memory for a sentence that conveyed much the same meaning in a figurative form (e.g., *"You can lead a horse to water, but you can't make it drink."*), then differences can be taken as evidence for differences in the processes underlying comprehension of figurative and literal language.

Three variants of this episodic approach are frequently found in the literature. With *recognition* memory, the participant has to identify whether a given statement is exactly the same as the earlier presented target. One can manipulate the distracter items (i.e., items not read earlier) along any of a number of dimensions. For instance, some distracters may have the same meaning as the target meaning, but the form in which they are presented can be altered (e.g., from active to passive or from literal to figurative). Falsely recognizing these distracters as having been read before is taken as evidence that the changed feature is not an important characteristic of the encoded target. Thus, falsely recognizing a literal translation of a metaphor as having been presented earlier is evidence that the meaning (which has not been altered) is encoded but that the form of presentation (which has been altered) is not retained.

A second episodic measure comes from *cued recall* studies. The same basic study procedure is employed under all conditions. When memory for the episode is tested, participants are given a hint ("cue"), and their task is to recall the target information. Some theories of memory claim that a retrieval cue is effective only if it was part of the encoding context; thus, one can get insights into processes that occurred during comprehension by examining which cues are effective and which are not. As an example, consider an early influential study by Verbrugge and McCarrell (1977). They found that the implicit resemblance (or ground) of a metaphor was a highly effective cue, suggesting that "subjects infer a resemblance during their initial encounter with a metaphoric sentence and that the resemblance is integral to what is stored as a memory of that experience" (p. 505).

Finally, one can measure episodic memory by a *free recall* procedure, in which participants reproduce the presented information in any order or way that comes to mind. Insights into the comprehension of figurative language can be found experimentally by manipulating the context in which a given target appears (e.g., Marschark & Hunt, 1985) or, using correlational techniques, by examining differences in the nature of the target (e.g., Kreuz, Long, & Church, 1991). For instance, Kreuz et al. found higher recall for ironic statements than for literal analogues of the irony. More important, the level of recall for both literal and ironic counterparts was predicted by the presence of pragmatic goals (e.g., showing anger, saving face, etc.). Taken together, these data suggest that the interpretation of speaker intent plays an important role in memory (and by implication, comprehension). These goals can be met by both literal and figurative statements but are more typically fulfilled by the ironic version.

Online Techniques

The techniques just described measure the result (on memory) established after a person has comprehended language. Memory and comprehension, although related, are not synonymous. It is easy to demonstrate that the two are dissociable: A person can remember a "nonsense" letter string such as a word in a foreign language without understanding what the word means, and Korsakoff amnesiacs can contribute meaningfully to a conversation but can forget even having the conversation within minutes. As such, memory-based measures are, at best, indirect indexes of the comprehension process and are subject to interpretive difficulty. A memory trace may be a result of either early or late processing of an input.

Online measures are intended to track mental processing while the processing occurs. There are an increasing number of online measures available for use. These measures typically depend on assumptions about the resources available for processing information: Difficult-to-process input is assumed to require more resources than does easy-to-process input and, consequently, should be done less efficiently. Moreover, because resources are limited, when one is processing difficult input, scant resources remain for simultaneously processing other input; thus, performance on secondary tasks should be performed less efficiently with difficult-to-process information.

Latency measures have been the preferred measure of processing efficiency. Some of these online measures commonly used (or available for use) in the figurative language literature follow.

Reading time. In this task, linguistic information is typically presented on a computer screen one unit (word, phrase, etc.) at a time, and the reader advances the unit at his or her own rate. Reading time is taken as an index of difficulty in processing the unit, with longer reading times indicating more difficulty. For example, Ortony, Schallert, Reynolds, and Antos (1978) measured the latency taken to read target sentences that were placed in contexts intended to give the target either a literal or figurative interpretation. Because figurative language has been assumed by some to involve more processing than is found with literal language, one might expect longer reading time for the sentences used figuratively. The failure to find latency differences between these two conditions was taken as disconfirmation of one of the major tenets of the standard pragmatic theory. Although reading time for complete sentences is the most popular measure found in figurative language research, the technique has been criticized as being fairly insensitive to moment-by-moment processing differences; consequently, in some studies, the RT for each word in the trope is analyzed. The other techniques that follow have not been employed as frequently in the figurative language literature and, in some cases, are limited, as is reading time, to visually presented stimuli.

Eye movements. Arguably this technique is becoming the procedure of preference in the reading literature because it is a naturalistic means of gathering online information. The popularity of this technique has been limited by the availability of the sophisticated equipment necessary for conducting eye tracking research, although increasingly more laboratories have acquired this tool. With this procedure, verbal material is presented and eye movements recorded such that one can trace which word is being examined (and for how long) at any given moment during the time spent reading. For example, Titone and Connine (1994), using eye tracking, showed that both meanings of an idiomatic expression are automatically available regardless of contextual bias.

Probe reaction time. A dual-task procedure is employed in which the participant is asked to perform a primary task (such as reading a passage, advancing units as described previously, etc.) while simultaneously performing a secondary task (such as pressing a key as soon as a certain sound, the probe, is presented). The experimenter can ensure that the probe is yoked to the primary task such that it appears simultaneously with theoretically interesting, primary-task information (e.g., the onset of an idiom or other trope). The assumption is that, as processing the primary task becomes more difficult, fewer resources are available to process the probe, leading to a slower reaction time on the probe task.

Phoneme monitoring task. In this task, participants listen to a steady stream of speech and are asked to press a key as soon as they hear a certain phoneme (e.g., the sound *ch*). The logic is identical to that of the probe reaction time task in that the participant is performing two tasks at the same time—comprehending the speech and monitoring a certain sound. As comprehension difficulty increases, there are less resources available for monitoring, and reaction time to the sound identification increases. A useful feature of this procedure is that it allows for online examination of spoken language.

Priming tasks. Processing of an input is made easier if information relevant to the input has already been accessed. Consider a word such as *bat.* If a person has processed that word (the "prime") and is immediately afterwards presented a related word, the related word (or "target") should be processed more rapidly (as evidenced by tasks such as the speed of reading the target or of deciding whether a letter string target is a real word). One can make inferences about the type of information made available online by manipulating the prime-target relation. For instance, consider reading "my favorite mammal is the bat," followed immediately afterward by the target *baseball* or the target *cave.* One would typically find a processing advantage (relative to an unrelated word control) for the target *cave,* indicating that processing the word *bat* brought related information to mind online. A question of interest is whether context-unrelated information also comes to mind. Finding a processing advantage for the target *baseball* indicates that multiple meanings of the word *bat* are accessed online, even if both are not consistent with the context.

One application of this logic to the study of figurative language has been to see if both the literal and figurative sense of a word or phrase come to mind when initially processed. Two variants of this task are frequently observed in the literature: *cross-modal priming,* in which the prime is presented as spoken speech and the target is presented visually, and a *repetition prime,* in which a word is visually presented in some text, the screen goes blank, and the word is presented again, with the task being to state whether the test word had been presented somewhere in the text. This latter technique, developed by Dell, McKoon, and Ratcliff (1983) has been used to study anaphor. For instance, the word *burglar* might be presented in the text, whereas later in the text, the word *criminal* is presented, and the test word that follows is *burglar.* The question of interest here is whether reading the anaphor *criminal* initiates an online search and activation in memory for the earlier reference (e.g., burglar). Such a reactivation would result in faster verification of the test word relative to an appropriate control item.

SELECTED RESEARCH QUESTIONS OF INTEREST

Research in figurative language has become increasingly more central to discussions of language in the 17 years since Ortony, Reynolds, and Arter (1978) published their seminal review paper. Selected aspects of this literature will be briefly reviewed.

Why Do We Use Metaphor and Other Forms of Figurative Language?

The basic issue is why someone would prefer to make an indirect request ("Do you have a dime?") rather than a direct request ("Would you lend me a dime?") or use a metaphor *("His car is a lemon")* rather than a literal equivalent *(His car is often in the garage)*. Wheelwright (1962), Ortony (1975), and others have made the claim that at least some instances of metaphor can serve an expository function not served by literal language, especially when dealing with creative thought. From this perspective, metaphor serves a function less easily expressed, or even inexpressible, in literal language (see Crider & Cirillo, 1989; Olson, 1988). Other functions of metaphor and other tropes have revolved around the social functions of figurative language use (see Crider & Cirillo, 1989; Furnham, 1987; Gerrig & Gibbs, 1988). From this perspective, figurative language serves to express socially relevant information such as intimacy, attributions, attitudes, and beliefs and may interact with metacognition about the vividness or persuasiveness of different types of messages.

Only recently have some of these claims been examined empirically. One set of studies examined the claim that metaphor serves a creative function. For instance, Kelly and Keil (1987) asked participants to rate a set of nouns along a set of semantic differential scales. For instance, some of the nouns could refer to magazines (*Sports Illustrated, The New Yorker,* etc.), others to foods (e.g., pizza, quiche). Both the control and experimental groups performed this task twice. From the ratings, similarity indexes could be formed between, for instance, *Sports Illustrated* and pizza. Between completion of the two sets of ratings, the experimental, but not the control, group was required to comprehend a metaphor created by pairing the terms (e.g., *"Sports Illustrated is the pizza of magazines")*. Kelly and Keil found that the concepts paired in the experimental group were understood as more similar than the corresponding pairs in the control group, suggesting that the act of metaphorization creates similarity between concepts. A similar conclusion was drawn by Camac and Glucksberg (1984), who demonstrated that words taken from apt metaphors did not produce any more priming than that found with randomly paired words and much less priming than that found with associatively related pairs. Tourangeau and Rips (1991) directly examined the features generated separately to the topic (such as *Sports Illustrated*) and vehicle (such as pizza) and to the metaphor as a whole. Many of the features of the metaphor were emergent, as confirmed by later ratings of the aptness of the metaphor.

There was a second, important aspect of Kelly and Keil (1987) that has not undergone further testing. They demonstrated that similarity was increased for words that came from the same conceptual domains as the words presented in a metaphor form, even though the specific words had not been presented (thus, terms like *The New Yorker* and *quiche* would be seen as more similar in the experimental group, even though participants would never had encountered the sentence *"The New Yorker is the quiche of magazines").* These data suggest that metaphor serves not only to reorganize specific concepts presented in the metaphor but also "to rework or reorganize a person's knowledge of the world" (Olson, 1988, p. 222).

A second set of studies examined the conditions under which one linguistic form or a specific metaphor is preferred to another. Edwards and Clevenger (1990), Gibb and Wales (1990), and Katz (1989), have examined production tasks in which participants are asked to complete a sentence. The preference for using specific linguistic forms such as sarcasm or making specific comparisons are then related to factors such as the nature of the context and characteristics of the choice (such as concreteness and affect). The results are quite variable, and more research along these lines is required. Gibbs and Mueller (1988) present an elegant set of studies on the use of indirect requests that can serve as a model for additional research. Using both production and judgment tasks, they demonstrated that speakers are sensitive to potential obstacles in achieving the end being requested, that these obstacles differ as a function of social context (also see Gibbs, 1986), and that speakers organize their conversations to deal with the relevant obstacles at hand.

What Is Represented?

A basic problem in cognitive science is in determining how the meaning of a trope is derived from the meaning of its parts. The answer to this question is, in part, constrained by the mental representation assumed to be underlying the trope. In this section, various representational approaches will be briefly addressed.

Is Mental Imagery Important?

The theoretical contrast is between representations based on amodal, abstract knowledge structures such as propositions and those based on perceptual experience, which is sometimes reflected in the conscious experience of imagery. The importance of imagery is somewhat uncertain. It is clear that participants can reliably judge the imagery-evoking characteristics of metaphor (e.g., Katz et al., 1988), proverbs (e.g., Benjafield, Frommhold, Keenan, Muckenheim, & Mueller, 1993), and other tropes. Moreover, these ratings are predictive of metaphor aptness (e.g., Katz, Paivio, & Marschark, 1985), episodic recall of metaphor (e.g., Marschark & Hunt, 1985), and choice of vehicle in a metaphor completion task (e.g., Katz, 1989). Paivio and Clark (1986) showed that presenting high imagery-evoking vehicle primes shortened the time necessary to interpret a metaphor using that vehicle, suggesting that imagery may work online.

In contrast to these positive results, other studies have argued against a role for imagery. Katz (1989), for instance, argued that the word-concreteness effects that he found were a function of the way that concrete words are represented in semantic memory and not of imagery per se. Katz based this conclusion on the relations between task performance and individual differences: Concreteness effects were found to vary with analogic reasoning ability but not with individual differences in imagery.

Other studies have more directly addressed the question by manipulating variables that presumably activate imagery to differing extents. Riechmann and Coste (1980) reviewed studies that indicate instructions to use imagery during comprehension hinder (and not, as predicted from the imagery perspective, facilitate) the comprehension of proverbs. Honeck (e.g., Feldhaus & Honeck, 1989; Honeck, Kibler, & Sugar, 1985) presented evidence that proverbs are encoded in an amodal conceptual format and that in judgment or memory tasks they are not understood in terms of imagery associated with the proverb or its instances. Under this view, however, imagery can play a role in facilitating the construction of figurative meaning. Fainsilber and Kogan (1984) demonstrated that rated quality of metaphor is independent of imagery ratings.

As is clear, a consensus on the role of imagery has not yet been established, and additional research on this important topic is expected.

What Must Be Mapped?

This controversy is a difference of opinion in what is assumed to constitute the essential characteristics of metaphor. Several distinctive classes of models are popular.

Multidimensional representational models. Based on multidimensional space models of similarity (see Katz, 1992, for a review), the most detailed model, by Tourangeau and Sternberg (1981), proposed that on presentation of a metaphor (e.g., "*Clinton is a pizza*"), higher order domains get engaged (e.g., politicians, foods), and a vector is constructed from the instance in the source domain to the parallel space in the target domain. In essence, one constructs an analogy in which one attempts to identify the characteristics of Clinton as a politician that parallels the characteristics of pizza to foods. The model predicts that the ease of comprehension depends on the similarity of characteristics shared by the domains and that aptness depends on the dissimilarity in characteristics of the domains themselves. The model has been tested by constructing stimulus materials that differ in the ways predicted by the model and obtaining judgments of comprehension and aptness, with results that correspond generally but weakly to the prediction (see Tourangeau & Sternberg, 1981; Trick & Katz, 1986; see also Katz, 1989).

Salience imbalance model. Ortony (1979) proposed that the representational characteristics of importance reside with the specific features that reside with the topic (e.g., Clinton) and vehicle (e.g., pizza) of a metaphor (*"Clinton is the pizza of politicians"*). The important featural characteristic is salience, defined relative to the term to which it is associated. The meaning and existence of a metaphor is based on features shared by the topic and vehicle in which those features are of high salience with respect to the vehicle but are of low salience with respect to the topic. For example, consider the metaphor *"Juliet is the sun."* The properties strongly associated with the sun (such as being central, life-giving, and warm) will not be salient descriptors of a person, although they can be employed in that way. Literal statements are, it is argued, not marked by such imbalance in saliency. The model can explain the asymmetry that exists when topics and vehicles are reversed (e.g., *"Pizza is the Clinton of foods"*). Using a set of judgment tasks, Ortony, Vondruska, Foss, and Jones (1985) provided empirical evidence that saliency differences occur in the predicted manner and that, as predicted, differences in saliency were greater for metaphor than for simile and literal statements alike. Recently, however, some have criticized the salience-imbalance position because, they argue, the distinction between literal and nonliteral statements is independent of the distinction between symmetric and asymmetric comparisons (see, for instance, Glucksberg & Keysar, 1990). Literal statements can demonstrate asymmetry when subject and predicate are reversed, and more important metaphorical comparisons can be symmetrical, as in those that occur when one compares the concepts *coal* and *night* (i.e., *"The night is like coal"* versus *"Coal is like the night"*).

Structure-mapping model. This model, which has been promoted by Gentner (e.g., Gentner & Clements, 1988) assumes a propositional base with structurally different components. The major assumptions are that one maps from vehicle to topic domains in ways that maintain systems of relations and that the choice of which relations to map is governed by higher order constraints—what Gentner calls the *principle of systematicity.* This model has been tested by having participants write descriptions of topic and vehicle and then write out interpretations when the terms are combined as metaphor. Featural and relational information was gleamed from the descriptions and used to predict judgments of metaphoricity and aptness. In general, the data were more supportive of structure-mapping than of salience-imbalance representational assumptions. Converging evidence for this general approach is also available from studies in which participants are asked to solve random generated analogies. In four studies, Johnson and Henley (1992) analyzed the reasons that participants used to generate their solutions to this unstructured task and found that they employed a small number of basic relational concepts. Exemplars include relations such as "goes to," "in," "on," and "part of."

None of the various representational schemes reviewed are, by themselves, sufficient to account for the wide range of tropes that one encounters. One can expect further tests of the extant models and the development of more sophisticated models in the future.

How Much Do the Components of a Trope Contribute to Overall Meaning?

Can the figurative interpretation of a topic be derived compositionally from the interpretation of its parts? Most of the research in this topic has involved idioms, many of which cannot be understood by understanding their component words (consider for instance *"kick the bucket"*). The naive argument is that such items are lexicalized, that is, represented in a manner similar to that for words. However, memory-based and online techniques have shown that the representation of idioms is more complex than suggested by the lexicalization hypothesis. The number of idiomatic meanings (Mueller & Gibbs, 1987), literalness, frequency of usage and frequency of the individual words (e.g., Cronk, Lima, & Schweigert, 1993), syntactic frozeness (e.g., Connine, Blasko, Brandt, & Layer, 1992), degree of semantic decomposition (Gibbs, Nayak, & Cutting, 1989), and other variables can be shown to influence the processing of idioms.

Because the meaning of an idiom such as *"flip one's lid"* cannot be determined from analysis of the constituent words, traditional theories hold that idiomatic meaning is arbitrary, and although they may have at one time been based on metaphor, over time the metaphorical sense has been lost. Gibbs (e.g., 1992), in contrast, argued that idioms should not be considered dead metaphors, because they have complex meanings that are motivated by existing conceptual metaphors. For instance, the idiom *"flip one's lid"* can be said to be motivated by two root metaphors, namely *"THE MIND IS A CONTAINER"* and *"ANGER IS HEATED FLUID IN A CONTAINER."* Gibbs and O'Brien (1990) demonstrated that people's mental images of a set of idioms based on the same root metaphors share constraints imposed by the root metaphors. This conclusion, based on imagery or judgment tasks for idioms placed in different scenarios, cannot, however, be taken as evidence that the presumed conceptual metaphors act online.

The number of variables now shown to have an effect indicates that the representational nature of idioms and, by implication, other forms of figurative language is extremely complex and is probably influenced by experiential factors (such as familiarity), syntactic factors (such as those that underlie the maintenance of idiomatic meaning as word order changes), the ease with which one can get to a figurative interpretation from the meaning of the parts, and the ease with which idioms can be mapped to larger conceptual structures. A more fundamental issue is whether these structures are inherently metaphorical as Gibbs (1994) and his colleagues claim, or whether nonmetaphorical structures and inferential processes will suffice as others have claimed (e.g., Glucksberg, Brown, & McGlone, 1993; Honeck & Temple, 1994).

What Is Processed?

The literature has been dominated by Searle's (1979) position that the processing of figurative language is triggered by a failure to find a suitable, context-appropriate literal interpretation. This theoretical position assumes, incorrectly, that an accepted definition of literal meaning exists (see, for instance, Dascal, 1987; Gibbs, 1989). Nonetheless, based on the notion that literal meaning is tied more directly to the referents being discussed, two implications of what some have labeled the *standard pragmatic model* have been tested extensively. These implications are that the processing of figurative meaning (a) is optional, and (b) should be computed only after a literal meaning is computed. A review of much of this literature can be found in Glucksberg (1991).

Research over the past 17 years has challenged both of these assumptions. Ortony, Schallert, Reynolds, and Antos (1978), Inhoff, Lima, and Carroll (1984), Shinjo and Myers (1987), and others have shown that metaphors are read as rapidly as literal translations when the items are placed in sufficiently elaborated discourse contexts. Similar findings have been found with other figures of language such as proverbs (Kemper, 1981) and idioms (Gibbs, 1980). These data indicate that figurative language can be understood at least as rapidly as literal language and disconfirms the second assumption that literal language has a processing priority. The first assumption, that figurative meaning is only processed optionally, has also been discredited. The evidence here is based on Stroop-like tasks: The argument is that the figurative sense of a trope should not interfere with the processing of the literal sense of the trope unless the figurative meaning is automatically (nonoptionally) aroused at about the same time that the literal meaning is aroused. However, interference effects have been found both with familiar metaphors (e.g., Glucksberg, Gildea, & Bookin, 1982) and with context-dependent items in which the metaphoricity of the target item is determined by the discourse (Keysar, 1989).

Two streams of research appear to be emerging from this line of work. The first addresses the problem of processing: If literal and figurative language are fundamentally the same, how then does one compute the nonliteral sense of a trope? A popular view is that contextual information plays an important role for both literal and metaphoric language. Gildea and Glucksberg (1983) manipulated the type of prime presented prior to metaphor comprehension and concluded that any context that activates properties of a predicate that are informative about the subject is sufficient to trigger an immediate interpretation. Thus, contextual primes relevant to a metaphoric vehicle that provide information about a metaphoric topic will be sufficient to trigger immediate interpretation. Others have presented evidence that, for a range of tropes, both figurative and literal senses of a sentence are processed in parallel if both are contextually plausible (e.g., Keysar, 1989; Titone & Connine, 1994). Others still have provided online evidence that only one interpretation is computed, although there is no preference for literal over figurative meaning in determining which interpretation is derived; presumably, context effects can act to bias one sense over another. In line with this possibility, manipulating instructions or blocks of trials so that a participant can expect figurative speech influences the speed and type of processing in which participants engage (see Alonzo-Quecuty & de Vega, 1991; Gregory, 1993).

The second, and not unrelated, line of research has been to show that a blanket dismissal of the standard pragmatic model is unwarranted. Most of the data taken as evidence against the model are based on familiar or conventional tropes. There is emerging an ever-increasing set of studies indicating that familiarity and other variables play an important role in determining when and if literal meaning is given processing priority (e.g., Blasko & Connine, 1993; Turner, 1995; Wolff & Gentner, 1992). Indeed, some recent computer models of metaphor processing have built in this distinction (e.g., Martin, 1992). Moreover, there is research indicating that even for familiar tropes, some processing of the literal sense can occur. For example, there is evidence that some literal processing occurs even for idioms (Needham, 1992) and for single words within idioms (e.g., Gibbs, 1990), although the extent of such processing is uncertain. Also, cross-modal priming studies indicate that if a word string is recognized as being an idiom early in the processing sequence, then literal processing might terminate at that point (see Cacciari & Tabossi, 1988). This approach assumes that one first attempts to process language at the most conventional level, which, for most language, will be based on literal meaning. Of course, this approach raises the question of how one recognizes that a trope is intended as a trope.

It is probable that at least some of this recognition will be determined by contextual and sentential factors such as those discussed earlier. For a trope to be recognized as a trope, the intended meaning will have to somehow be marked in the discourse context. Onishi and Murphy (1993) argued that marking may be one way that literal and figurative language differ. That is, with literal language, it may be easier to find coherence in discourse due to semantic similarity. For instance, if one talks about "boxers" at a fighting match and later mentions the word *fighter,* the connection between the current use of *fighter* and its earlier reference may be more transparent than if the later usage was metaphoric (e.g., *"the cream-puff"*). If so, a difference in literal and figurative language would emerge online as the figurative expression became progressively less marked in context.

Do Conceptual Structures Play an Active Role in Processing?

Lakoff and Johnson (1980) provided evidence that our understanding of the world is motivated by a set of root metaphors such as *"THE MIND IS A CONTAINER."* The evidence is based on the systematicity of many forms of expression: Diverse expressions often can be grouped under a metaphor in which one domain of experience (such as those dealing with the *"MIND"*) can be understood in terms of a very different (usually concrete) domain (such as *"CONTAINERS"*). For instance, people can be said to *"have closed (or open) minds," "to blow their tops," "to be empty headed,"* and so forth. Thus, assuming the existence of root metaphors makes it possible to explain the relations found between seemingly unrelated sentences.

Gibbs (e.g., see Gibbs & O'Brien, 1990; Nayak & Gibbs, 1990, for an analysis of idioms) has been the most vocal in arguing that these root metaphors constrain our understanding of figurative (and literal) expressions and, hence, play an active role in when and how we use the expressions and in the on-going processing of these expressions. The supporting data comes from both judgment tasks and reaction-time measures.

Gibbs's position is quite controversial and has led to a vigorous debate in the literature (e.g., Gibbs, 1992; Gibbs & Nayak, 1991; Glucksberg, Keysar, & McGlone, 1992; Honeck & Temple 1994; Kreuz and Graesser, 1991). An elegant experimental rebuttal is provided by Glucksberg et al. (1993). They found some evidence in favor of conceptual metaphors using judgment tasks, a finding consistent with those reported by Gibbs. However, online reading time measures indicated no advantage for idioms that were consistent with an underlying conceptual metaphor when compared to idioms that were analogically inconsistent. Taken together, these data suggest that conceptual metaphors are not automatically engaged during comprehension, although they may play a role under appropriate conditions. These conditions, if they exist, remain to be identified.

SUMMARY

The experimental psycholinguistic literature on figurative language has matured greatly in the 17 years since Ortony, Reynolds, and Arter (1978) published their review, both methodologically and conceptually. Today there are numerous well-controlled and conceptualized studies not only on metaphor but on indirect requests, idioms, sarcasm, and many other figurative forms. Methodological advances have permitted a finer analysis not only of the integration of tropes that results from the act of comprehension but also of the online processing that occurs during the act itself. There is also an ongoing attempt to integrate the use of given expressions within a larger social and conceptual framework than that afforded by current computational models.

As this brief review indicates, much remains to be done. The pragmatic functions of various expressions are not well understood, and there is scant research dealing with the effects of using figurative language (rather than a literal translation) or of one form of expression rather than another. Although, compared with the research described in 1978, there are many more studies that employ context-based stimuli, the materials found in the current experimental literature are still, for the most part, overly simple, nonrealistic, and presented independently of the social contexts in which language is commonly used. The representational question has, for the most part, been considered independently of pragmatic issues, but almost certainly how a trope is represented will depend on how it is to be used. Moreover, the various representational schemes just outlined are quite specific, and none can handle the wide range of tropes that are experienced daily. Finally, the role of affect must be clarified.

Current representational and processing models are best thought of as functional models: They outline the type of constraints that must be considered without detailing the algorithm by which these constraints can be implemented (although limited computational models are becoming available in the computer science literature). To be psychologically real, process models will have to incorporate variables known to influence processing such as familiarity, while simultaneously being consistent with more general psycholinguistic models such as those attempting to understand bridging inferences, thematic roles, and so forth. Finally, the roles played by context, that provided by the discourse and that brought to the communication situation, has to be better understood. In recent years, contextual effects such as those provided by common ground (e.g., Gerrig & Litman, 1990) or authorial intention (e.g. Gibbs, Kushner, & Mills, 1991), have been experimentally demonstrated. The participatory and emotional effects generated by context have now been studied experimentally but remain to be explored in depth (see Allbritton & Gerrig, 1991). One possibility is that such factors play an important role in inviting different interpretations of the same statement such as that which often occurs with sarcasm (see Katz, in press). Much of this kind of work should appear within the next few years.

ACKNOWLEDGMENT

Preparation of this article was supported by Grant OGP007040 from the Natural Sciences and Engineering Research Council of Canada.

REFERENCES

Allbritton, D., & Gerrig, R. (1991). Participatory responses in text understanding. *Journal of Memory and Language, 30,* 603–626.

Alonso-Quecuty, M., & de Vega, M. (1991). Contextual effects in a metaphor verification task. *European Journal of Cognitive Psychology, 3,* 315–341.

Benjafield, J., Frommhold, K., Keenan, T., Muckenheim, R., & Mueller, D. (1993). Imagery, concreteness, goodness, and familiarity ratings for 500 proverbs sampled from the Oxford Dictionary of English Proverbs. *Behavior Research Methods, Instruments, and Computers, 25,* 27–40.

Blasko, D., & Connine, C. (1993). Effects of familiarity and aptness on metaphor processing. *Journal of Experimental Psychology: Learning, Memory, and Cognition, 19,* 295–308.

Cacciari, C., & Tabossi, P. (1988). The comprehension of idioms. *Journal of Memory and Language, 27,* 668–683.

Camac, M., & Glucksberg, S. (1984). Metaphors do not use associations between concepts, they are used to create them. *Journal of Psycholinguistic Research, 13,* 443–455.

Connine, C., Blasko, D., Brandt, R., & Layer, J. (1992). Idiomatic processing: Syntactic frozenness and subjective familiarity. *Psychological Research, 54,* 225–232.

Crider, C., & Cirillo, L. (1989). Systems of interpretation and the function of metaphor. *Journal for the Theory of Social Behavior, 21,* 171–195.

Cronk, B., Lima, S., & Schweigert, W. (1993). Idioms in sentences: Effects of frequency, literalness and familiarity. *Journal of Psycholinguistic Research, 22,* 59–82.

Dascal, M. (1987). Defending literal meaning. *Cognitive Science, 11,* 259–281.

Dell, G., McKoon, G., & Ratcliff, R. (1983). The activation of antecedent information during the processing of anaphoric reference in reading. *Journal of Verbal Learning and Verbal Behavior, 22,* 121–132.

Edwards, R., & Clevenger, T. (1990). The effects of schematic and affective processes on metaphoric invention. *Journal of Psycholinguistic Research, 19,* 91–102.

Fainsilber, L., & Kogan, N. (1984). Does imagery contribute to metaphoric quality? *Journal of Psycholinguistic Research, 13,* 383–391.

Feldhaus, R., & Honeck, R. (1989). The conceptual basis of graded categories. *Journal of Psycholinguistic Research, 18,* 271–288.

Furnham, A. (1987). The proverbial truth: Contextually reconciling the truthfulness of antonymous proverbs. *Journal of Language and Social Psychology, 6,* 49–55.

Gentner, D., & Clements, C. (1988). Evidence for relational selectivity in interpreting analogy and metaphor. In G. Bower (Ed.), *The psychology of learning and motivation,* Vol. 22 (pp. 307–358). New York: Academic.

Gerrig, R., & Gibbs, R. (1988). Beyond the lexicon: Creativity in language production. *Metaphor and Symbolic Activity, 3,* 1–19.

Gerrig, R., & Litman, M. (1990). Disambiguation by community membership. *Memory and Cognition, 18,* 331–338.

Gibb, H., & Wales, R. (1990). Metaphor or simile: Psychological determinants of the differential use of each sentence form. *Metaphor and Symbolic Activity, 5,* 199–213.

Gibbs, R. (1980). Spilling the beans on the understanding and memory for idioms in conversation. *Memory and Cognition, 8,* 449–456.
Gibbs, R. (1986). What makes some indirect speech acts conventional? *Journal of Memory and Language, 25,* 181–196.
Gibbs, R. (1989). Understanding and literal meaning. *Cognitive Science, 13,* 243–251.
Gibbs, R. (1990). Comprehending figurative referential descriptions. *Journal of Experimental Psychology: Learning, Memory, and Cognition, 16,* 56–66.
Gibbs, R. (1992). Categorization and metaphor understanding. *Psychological Review, 99,* 572–577.
Gibbs, R. (1994). *The poetics of mind.* Cambridge, England: Cambridge University Press.
Gibbs, R., Kushner, J., & Mills, W. (1991). Authorial intentions and metaphor comprehension. *Journal of Psycholinguistic Research, 20,* 11–30.
Gibbs, R., & Mueller, R. (1988). Conversational sequences and preference for indirect speech acts. *Discourse Processes, 11,* 101–116.
Gibbs, R., & Nayak, N. (1991). Why idioms mean what they do. *Journal of Experimental Psychology: General, 120,* 93–95.
Gibbs, R., Nayak, N., & Cutting, C. (1989). How to kick the bucket and not decompose: Analyzability and idiom processing. *Journal of Memory and Language, 28,* 576–593.
Gibbs, R., & O'Brien, J. (1990). Idioms and mental imagery: The metaphorical motivation for idiomatic meaning. *Cognition, 36,* 35–68.
Gildea, P., & Glucksberg, S. (1983). On understanding metaphor: The role of context. *Journal of Verbal Learning and Verbal Behavior, 22,* 577–590.
Glucksberg, S. (1991). Beyond literal meanings: The psychology of allusion. *Psychological Science, 2,* 146–152.
Glucksberg, S., Brown, M., & McGlone, M. (1993). Conceptual metaphors are not automatically accessed during idiom comprehension. *Memory and Cognition, 21,* 711–719.
Glucksberg, S., Gildea, A., & Bookin, H. (1982). On understanding nonliteral speech: Can people ignore metaphors? *Journal of Verbal Learning and Verbal Behavior, 21,* 85–98.
Glucksberg, S., & Keysar, B. (1990). Understanding metaphorical comparisons: Beyond similarity. *Psychological Review, 97,* 3–18.
Glucksberg, S., Keysar, B., & McGlone, M. (1992). Metaphor understanding and accessing conceptual schema: Reply to Gibbs (1992). *Psychological Review, 99,* 578–581.
Gregory, M. (1993). Metaphor comprehension: From literal truth to metaphoricity and back again. *Metaphor and Symbolic Activity, 8,* 1–21.
Honeck, R., Kibler, C., & Sugar, J. (1985). The conceptual base view of categorization. *Journal of Psycholinguistic Research, 14,* 155–174.
Honeck, R., & Temple, J. (1994). Proverbs: The extended conceptual base and great chain metaphor theories. *Metaphor and Symbolic Activity, 9,* 85–112.
Inhoff, A., Lima, S., & Carroll, P. (1984). Contextual effects on metaphor comprehension in reading. *Memory and Cognition, 12,* 558–567.
Johnson, M., & Henley, T. (1992). Finding meaning in random analogies. *Metaphor and Symbolic Activity, 7,* 55–75.
Katz, A. (1982). Metaphoric relationships: The role of feature saliency. *Journal of Psycholinguistic Research, 11,* 283–296.
Katz, A. (1989). On choosing the vehicles of metaphors: Referential concreteness, semantic distances, and individual differences. *Journal of Memory and Language, 28,* 486–499.
Katz, A. (1992). Psychological studies in metaphor processing: Extensions to the placement of terms in semantic space. *Poetics Today, 13,* 607–632.
Katz, A. (in press). Pragmatics and the creation of similarity in metaphor and irony. In J. Mio & A. Katz (Eds.), *Metaphor: Implications and applications.* Mahwah, NJ: Lawrence Erlbaum Associates, Inc.

Katz, A., Paivio, A., & Marschark, M. (1985). Poetic comparisons: Psychological dimensions of metaphoric processing. *Journal of Psycholinguistic Research, 14,* 365–383.

Katz, A., Paivio, A., Marschark, M., & Clark, J. (1988). Norms for 204 literary and 260 nonliterary metaphors on 10 psychological dimensions. *Metaphor and Symbolic Activity, 3,* 191–214.

Kelly, M., & Keil, F. (1987). Metaphor comprehension and knowledge of semantic domains. *Metaphor and Symbolic Activity, 2,* 33–52.

Kemper, S. (1981). Comprehension and the interpretation of proverbs. *Journal of Psycholinguistic Research, 10,* 179–198.

Keysar, B. (1989). On the functional equivalence of literal and metaphorical interpretations in discourse. *Journal of Memory and Language, 28,* 375–385.

Kreuz, R., & Graesser, A. (1991). Aspects of idiom interpretation: Comment on Nayak and Gibbs. *Journal of Experimental Psychology: General, 120,* 90–92.

Kreuz, R., Long, D., & Church, M. (1991). On being ironic: Pragmatic and mnemonic implications. *Metaphor and Symbolic Activity, 6,* 149–162.

Lakoff, G., & Johnson, M. (1980). *Metaphors we live by.* Chicago: Chicago University Press.

Marschark, M., & Hunt, R. (1985). On memory for metaphor. *Memory and Cognition, 13,* 413–424.

Martin, J. (1992). Computer understanding of conventional metaphoric language. *Cognitive Science, 16,* 233–270.

Mueller, R., & Gibbs, R. (1987). Processing idioms with multiple meanings. *Journal of Psycholinguistic Research, 16,* 63–81.

Nayak, N., & Gibbs, R. (1990). Conceptual knowledge in the interpretation of idioms. *Journal of Experimental Psychology: General, 119,* 315–330.

Needham, W. (1992). Limits on literal processing during idiom interpretation. *Journal of Psycholinguistic Research, 21,* 1–16.

Olson, D. (1988). Or what's a metaphor for? *Metaphor and Symbolic Activity, 3,* 215–222.

Onishi, K., & Murphy, G. (1993). Metaphoric reference: When metaphors are not understood as easily as literal expressions. *Memory and Cognition, 21,* 763–772.

Ortony, A. (1975). Why metaphors are necessary and not just nice. *Educational Theory, 25,* 45–53.

Ortony, A. (1979). Beyond literal similarity. *Psychological Review, 86,* 161–180.

Ortony, A., Reynolds, R., & Arter, J. (1978). Metaphor: Theoretical and empirical research. *Psychological Bulletin, 85,* 919–943.

Ortony, A., Schallert, D., Reynolds, R., & Antos, S. (1978). Interpreting metaphors and idioms: Some effects of context on comprehension. *Journal of Verbal Learning and Verbal Behavior, 17,* 465–477.

Ortony, A., Vondruska, R., Foss, M., & Jones, L. (1985). Salience, similes, and the asymmetry of similarity. *Journal of Memory and Language, 24,* 569–594.

Paivio, A., & Clark, J. (1986). The role of topic and vehicle imagery in metaphor comprehension. *Communication and Cognition, 19,* 367–388.

Riechmann, P., & Coste, E. (1980). Mental imagery and the comprehension of figurative language: Is there a relationship? In R. Honeck & R. Hoffman (Eds.), *Cognition and figurative language* (pp. 183–200). Hillsdale, NJ: Lawrence Erlbaum Associates, Inc.

Searle, J. (1979). Metaphor. In A. Ortony (Ed.), *Metaphor and thought* (pp. 92–123). Cambridge, England: Cambridge University Press.

Shinjo, M., & Myers, J. (1987). The role of context on metaphor comprehension. *Journal of Memory and Language, 26,* 226–241.

Titone, D., & Connine, C. (1994, March). *Taking on semantic commitments: Processing semanti-*

cally non-decomposable and decomposable idioms. Poster presented at the 7th Annual CUNY Conference on Human Sentence Processing, New York.

Tourangeau, R., & Rips, L. (1991). Interpreting and evaluating metaphors. *Journal of Memory and Language, 30,* 452–472.

Tourangeau, R., & Sternberg, R. (1981). Aptness in metaphor. *Cognitive Psychology, 13,* 203–244.

Trick, L., & Katz, A. (1986). The domain interaction approach to metaphor processing: Relating individual differences and metaphor characteristics. *Metaphor and Symbolic Activity, 1,* 185–213.

Turner, N. (1995). *The role of literal meaning in proverb comprehension.* Unpublished doctoral dissertation, University of Western Ontario, London, Ontario, Canada.

Verbrugge, R., & McCarrell, N. (1977). Metaphoric comprehension: Studies in reminding and resembling. *Cognitive Psychology, 9,* 494–533.

Wheelwright, P. (1962). *Metaphor and reality.* Bloomington: Indiana University Press.

Wolff, P., & Gentner, D. (1992). The time course of metaphor comprehension. *Proceedings of the 14th Annual Conference of the Cognitive Science Society,* 504–509.

METAPHOR AND SYMBOLIC ACTIVITY *11*(1), 39–66
Copyright © 1996, Lawrence Erlbaum Associates, Inc.

On Perceptual Metaphors

Lawrence E. Marks
John B. Pierce Laboratory and Yale University

Metaphors reflect processes of thinking and, consequently, appear not just in language but in perception as well. A notable example of metaphorical perception is synesthesia, in which music or voices may be perceived to have shapes, textures, or colors. Another example is physiognomic perception, in which dark objects may be perceived as gloomy or bright ones may be perceived as happy. Synesthetic perception and physiognomic perception alike reveal the presence of deep similarities across different sense modalities, and many of these similarities appear to be inherent to perception. In this regard, intersensory and physiognomic metaphors reflect "natural" rather than "conventional" symbols or signs. Despite their perceptual origin, however, metaphorical relations evident in synesthetic and physiognomic perception are also represented in and accessible through language. Consequently, in information processing tasks requiring people to identify or categorize perceptual stimuli, linguistic or abstract processes, rather than strictly perceptual processes, may mediate cross-modal, metaphoric interactions.

Several major theses resound through the burgeoning literature on the psychology of metaphor. Among these is the growing recognition that metaphor is more than just a "rhetorical device" to embellish writing or speech, indeed that it is more than a linguistic phenomenon per se. Following hints from psychologists such as Arnheim (1969), literary critics such as Richards (1936), and philosophers such as Black (1962), it is now widely acknowledged that the processing of verbal metaphors differs in no deep way from that of literal statements (e.g., Gibbs, 1984; Glucksberg & Keysar, 1990) and that metaphorical processes are fundamental properties of cognition and conceptualization and not just of language (cf. Lakoff & Johnson, 1980). From this, it is a short step to infer that metaphor involves mechanisms of thinking and not just mechanisms of writing or speaking (e.g., Bowers, 1990; Haskell, 1987; Kennedy, 1990).

Requests for reprints should be sent to Lawrence E. Marks, John B. Pierce Laboratory, 290 Congress Avenue, New Haven, CT 06519. E-mail: marks@yalevm.cis.yale.edu.

In recent years, psychological investigations of metaphorization and figurative processes have broadened greatly in scope. As a consequence, even though the bulk of research into the processes of producing and understanding metaphor still focuses on the domain of figurative language, the recognition that figurative processes can transcend language has opened the study of metaphor to nonlinguistic forms.

PICTORIAL METAPHOR

More than a decade ago, Kennedy (1982) pointed out that pictures as well as words may be metaphorical or literal. In his view, the use of metaphor in pictures involves the intentional violation of a normal mode of generally recognized depiction (see also Kennedy, Green, & Vervaeke, 1993). In the course of indicating both the communalities and differences between pictorial and verbal figuration, Kennedy recounted the many analogues in pictures to standard forms of rhetorical speech such as allegory, catechresis, and hyperbole. Consider one of his examples: The rhetorical term *catechresis* refers to an expression that "fill[s] in a gap in the lexicon" (Kennedy, 1982, p. 594). A common example is the use of *fell* in the sentence "He fell in love." Lacking a word to express the thought underlying this utterance, the verb *to fall* is used metaphorically to refer to entering the state of being in love. It is a curious fact, perhaps a fact of language but maybe one of sensory psychology, that the sense of smell has virtually no vocabulary of its own, but instead expresses odor experiences through catechreses. The words we apply to smells represent either the objects that produce them metonymically, as in *rose,* or their qualities through analogues derived from other modalities, as in *bitter*.

Kennedy's (1982) pictorial analogue to catechresis is a set of line segments used in "inappropriate" places, typically in midair or space, to indicate motion or speed. This is a common pictorial device in comic strips and one that is readily understood immediately by all but the very young or individuals unfamiliar with pictorial displays (see Kennedy et al., 1993; in this regard, Stevenson & Friedman, 1986, pointed out that preschool children have particular difficulty understanding conventional pictorial representations of sound). As Kennedy noted, "speed lines" are sometimes ecologically literal, as when a car leaves tracks in dirt or snow. On the other hand, similar sets of lines can be used to express properties besides motion, as when lines drawn as if they emanate from a thumb indicate pain. Kennedy et al. (1993) also noted that blind individuals having little experience with pictures nevertheless often "figure out" what these pictorial metaphors mean in raised-line drawings.

Speed lines are generally treated within the context of representational drawings (pictures or cartoons). However, one can find similar groupings of lines in nonrepresentational works, for example, in abstract art. Many of the paintings of Kandinsky, especially those done in the 1920s and 1930s, contain small groups of lines, sometimes straight (a good example is *Thorough-Going Line* of 1923, with its multiple sets of parallel, converging, and diverging lines) and sometimes curved (as in *Little Dream in Red* of 1925). My own sense is that, there, too, these serve as speed lines, suggesting movement and temporal dynamism that reinforce the strong sense of musicality in Kandinsky's works.

As just indicated, the study of figurative processes in perception is predicated on the notion that it is thinking and not just talking or reading that is metaphorical. Even so, language is so very central to many forms of human thought. Recognizing the centrality of language to human cognition should, therefore, sensitize us to the possibility that verbal processes may play a role, perhaps minor but possibly major, even in the production or understanding of perceptual metaphors. Verbal mechanisms may be intimately involved in the ways that we decode, assess, formulate, and remember figures of thought, even when these are conveyed in perceptual rather than linguistic form. Later in this article, I take up the matter of how verbal codes might mediate certain metaphoric (cross-sensory and physiognomic) expressions.

It would be misleading to infer that research into nonlinguistic metaphor (an infelicitous but functional expression) is a wine of a wholly new vintage. As a matter of fact, some of its varieties—for instance, studies concerned with synesthesia—have had a relatively long history. Synesthesia may be deemed a kind of perceptual metaphor, in which qualities of one class or modality of sensory-perceptual experience are transferred to another, as when colors are ascribed to tastes, pains, or musical notes. To strongly synesthetic individuals, the transfer is typically asymmetric although the reasons for this asymmetry remain obscure. Sounds may take on colors, but colors may not have timbres or pitches. This asymmetry resembles the nonreflexivity typical of metaphorical language (the metaphor *"John is a fox"* cannot be reversed). Given the theoretical stance that treats metaphor as a mode of thinking, it is not a metaphorical ploy to call synesthesia a perceptual metaphor.

SYNESTHESIA AND CROSS-MODAL SIMILARITY

In treating it as metaphor, I am using the term *synesthesia* narrowly. Elsewhere (Marks, 1975, 1978), I have distinguished its strong and weak forms. Strong synesthesia is exceedingly rare, appearing in only a very small fraction of the population. To strongly synesthetic individuals, perceptual stimuli presented in one modality, typically hearing, produce qualitative experiences normally considered appropriate to a second modality, typically vision.

Thus, individuals who experience strong auditory–visual synesthesia may perceive music not only in terms of orchestral timbres and vocal melodies, but also as a dynamic display of shapes, contours, and colors. To strongly synesthetic perceivers, the intermodal experiences are unavoidable and ineluctable, with the correspondences between particular auditory qualities and particular visual qualities remaining stable over long periods of time.

Some investigators, such as Cytowic (1989), reserve the term *synesthesia* for individuals of this ilk, whose intermodal percepts may reflect discrete neurological processes involving regions such as the limbic system (or perhaps cerebral cortex; cf. Bar-Cohen, Harrison, Goldstein, & Wyke, 1993). From his neurological perspective, Cytowic notes five well-known characteristics of synesthesia that he considers to be diagnostic of its strong, idiopathic form: (a) Synesthesia is involuntary and is invariably elicited by a particular set or modality of stimuli pertinent to a given individual; (b) synesthetic experiences (particularly visual ones) are projected externally; (c) the cross-modal relations observed in synesthesia are, for a given individual, reliable and stable over long periods of time; (d) these experiences are vivid and easily remembered; and (e) strongly synesthetic individuals ascribe a validity to their experiences that Cytowic interprets as having an emotional basis. All of these characteristics are consistent with the view that strong synesthesia exemplifies perceptual metaphor and, in my view, is appropriately deemed to be cognitive in the broad sense of that term, although it surely need not involve language.

A couple of the characteristics of strong synesthesia also mark its weaker forms. Investigations over half a century ago (e.g., Karwoski & Odbert, 1938) noted that 10% to 15% of the population on occasion may experience a similar but much weaker translation of sensory qualities from modality to modality, often under voluntary control (for review, see Marks, 1975, 1978). Weaker still are those synesthetic tendencies evident in the vast majority of people: individuals who never report seeing voices or musical instruments as having colors, that is, who do not report that visual "sensations" or "images" accompany sounds but who nevertheless may describe a soprano's tone as lighter or sharper than an alto's, the register of a violin as lighter or sharper than a cello's, or a sneeze as brighter and sharper than a cough.

Is it useful to embrace with one term the entire range of phenomena, from strong to very weak (albeit common) synesthetic perception? Perhaps not, for such a classification may oversimplify and is likely to result in treating all of the manifestations as one and the same. That all synesthesias share certain properties—and, hence, may share underlying mechanisms of sensory-perceptual information processing—seems reasonably certain. But it is becoming increasingly apparent that the underlying processes also differ in part and, in particular, that cases of strong synesthesia probably have a special neurophysiological basis. Consequently, I shall refer to weak synesthetic tendencies as examples of cross-modal similarity.

Having said this, the resemblance between synesthesia in both its strong and weak forms and synesthetic tendencies expressed as cross-modal similarities is striking. Many of the common, central characteristics were first described in a series of papers by Karwoski, Odbert, and colleagues beginning in the 1930s (Karwoski & Odbert, 1938; Karwoski, Odbert, & Osgood, 1942; Odbert, Karwoski, & Eckerson, 1942; Riggs & Karwoski, 1934).

First and foremost, both synesthesia and cross-modal similarity consist, in large measure, of the alignment of perceptual *dimensions* in different modalities, for instance, of auditory pitch with visual brightness and lightness. Brightness refers to a dimension that applies to lights or levels of illumination and that runs perceptually from dim to bright, whereas lightness refers to a dimension that applies to reflecting surfaces and that runs perceptually from dark to light (from black to white if the surface is achromatic). Just as synesthetic perceivers "see" low-frequency sounds as dim or dark in color and high-frequency sounds as bright or light, so nonsynesthetic individuals judge low tones to be more like brown or black colors, high tones more like yellow or white. Similarly, low-frequency sounds express larger visual sizes than do high-frequency sounds, and soft sounds express dim visual colors, whereas loud sounds express bright ones. Relative to high-frequency sounds, low-frequency sounds also produce more rounded shapes, a visual attribute of form that is also connected to slower rather than faster musical tempos (see Marks, 1975, 1978).

A reasonably persuasive case can be made that several, although maybe not all, of the cross-sensory correspondences evident in synesthesia and in cross-modal similarity reflect hard-wired properties of perceptual systems. The strongest evidence comes from studies of infants and young children. Testing 3-week-old infants, Lewkowicz and Turkewitz (1980) found an absolute correspondence between brightness and loudness measured through generalization of cardiac deceleration. After exposure to a sound of given loudness, the infants' habituation in heart rate generalized to lights of equivalent brightness, this equivalence itself being gauged in cross-modality matches obtained in adults. Testing older infants, just under 1 year old, Wagner, Winner, Cicchetti, and Gardner (1981) reported "matches" between visual arrows pointing up or down and tones sweeping up or down in sound frequency. Testing still older children, Roffler and Butler (1968) reported that 4- to 5-year-olds perceived high-frequency sounds to be spatially higher than low-frequency tones. Marks, Hammeal, and Bornstein (1987) found that 4- to 5-year-olds reliably matched bright lights with loud and high-pitched sounds and dim lights with soft and low-pitched sounds.

Maurer and Maurer (1988) characterized early infancy as a period of synesthetic conflation of sensory experience—much like James's (1890) "blooming, buzzing confusion." More recently, Maurer (1993) suggested that the first month of life is marked by a kind of neonatal synesthesia in which babies endeavor to keep the total energy entering through the senses, regardless of sensory modality, within an optimal range (see Lawson & Turkewitz, 1980). Further, according to Maurer, this synesthetic integration over the senses tends to disappear soon in development as the various modalities become increasingly differentiated. By this token, it is tempting to reckon both the strong synesthesias evident in a small minority of people and the weaker, cross-modal similarities evident more pervasively in the vast majority as vestiges of an early, synesthetic state of development.

PHYSIOGNOMIC PERCEPTION

The position expressed by Maurer (1993) has deep roots. Feré (1892), more than a century ago, and Börnstein (1936), nearly half a century later, both offered a theory of "equivalent excitation," claiming that all sensory stimuli, regardless of the modality through which they are delivered, produce generalized physiological responses characterized by changes in such properties as muscle tonicity. Being undifferentiated, these responses can mediate synesthesia-like equivalences: Synesthetic communality between a high-pitched sound and a bright light occurs when the two stimuli evoke the same tonic motor response.

This theory received its fullest exposition in the writings of Werner (e.g., Werner, 1934, 1957, 1963; Werner & Wapner, 1949, 1952). Indeed, research conducted under the rubric of Werner and Wapner's sensory-tonic field theory demonstrated clear and potent "metaphorical" interactions among perceptual stimuli, muscle tone, and body orientation. For example, when participants try to set the orientation of a rod viewed in darkness to its "apparent vertical," the setting is systematically affected by asymmetric non-visual stimulation, for instance, by a sound stronger at one ear or by electrical stimulation delivered primarily to muscles on one side of the neck (Wapner, Werner, & Chandler, 1951). These interactions suggest a functional equivalence among the various sensory modalities together with the motor and proprioceptive systems.

One of the manifestations of such functional equivalence—according to Werner (1934, 1957), a manifestation directly connected to synesthesia—is the metaphorization known as *physiognomic perception.* Although, as Lindauer (1990a) noted, the word *physiognomy* commonly applies to expressive properties of the human body, especially the face, the term *physiognomic perception* has come to take on a broader meaning, referring more generally to dynamic, evocative, or expressive properties, for example, the appearance of a thundercloud as threatening, a skylark's song as joyful, the color black as somber, or the color yellow as happy. Physiognomic perception is "perception whereby the perceiver tends to suffuse percepts with an emotion, affective, or expressive quality" (Schlessinger, 1980, p. 74) replete with "nonliteral sensory, perceptual, and affective connotations" (Lindauer, 1990b, p. 47). According to Haskell (1987), Vico's theory of metaphor places its origin in physiognomic perception: "In some distant and dim past, says Vico, groups of men huddled together and *identified* a sudden and loud clap of thunder as *anger*. Thus was created the first metaphor" (Haskell, 1987, p. 69).

Within the visual realm, research into physiognomic or affective responses has focused largely on the properties of color and shape. The notion that color can affect a person's mood and behavior is widespread and has attracted attention in a variety of disciplines within psychology. For if colors have emotional connotations, then one might expect color to affect a wide range of everyday behaviors. For instance, a study by Damhorst and Reed (1986) reported that men (but not women) judged female job applicants to be more powerful if they wore dark-colored rather than light-colored jackets, and Frank and Gilovich (1988) found greater perceived and actual aggressiveness in football and hockey teams wearing black rather than white uniforms.

That dark and saturated colors are physiognomically stronger than light, pale colors is also evident in results obtained by D'Andrade and Egan (1974) and Valdez and Mehrabian (1994) using simpler color stimuli. D'Andrade and Egan presented to two groups of participants—one native speakers of English and the other native speakers of Tzeltal (a language of Mexico)—a series of words (in their native language) denoting various emotions, the task being to select for each emotion word the most appropriate color chip from a large array. Multidimensional scaling revealed striking similarities in the responses obtained from the two groups, especially with regard to the colors associated with "good," "bad," "strong," and "weak." To speakers of both English and Tzeltal, dark and highly saturated colors were good and strong, and light and palely saturated colors were weak. Negative emotions such as "sadness" and "fright," tended to be connected with purplish and yellow-red colors, and positive emotions such as "happy" were connected with blue-greens.

More recently, Valdez and Mehrabian (1994) had participants rate each of several colors on a variety of scales defining emotional qualities. Light and richly saturated colors were pleasant, whereas dark and saturated colors were arousing and dominating (strong). It is worth noting that variations of colors along the visual dimensions of brightness and saturation were generally much more salient and significant in evoking affective responses than was variation along the dimension of hue per se (see also Wright & Rainwater, 1962). Contrary to the popular notion that physiognomic responses are characteristic of particular hues themselves—for instance, that the color yellow is happy—it turns out that hue proper only weakly determines emotional response.

There is, I believe, a simple explanation for the apparent discrepancy, and it is connected to a confounding of dimensions of color percepts with color names. Consider the color name *yellow.* It refers not only to a particular class of hues with dominant wavelengths in the region of 570 nm but also to prototypical levels of brightness and saturation. Colors that we denote as yellow are, as a class, low in saturation and bright or light. Yellow, at its richest, is still the least saturated of hues. Where blues and greens and reds may be either richly saturated or pale, yellows inhabit the bottom of the scale of color saturation, never achieving high saturation. Yellow is always light and is intrinsically light: To dark yellows, we give different color labels, such as *tan.* By this token, it is its relatively great lightness and low saturation and not its hue per se that make yellow (metonymically) a "happier" color than blue.

In this there is an important underlying principle: In the assessment of figurative, metaphorical, synesthetic, or physiognomic properties of perception, it can be crucial to distinguish the perceptual experiences themselves from the linguistic labels that may become attached to them. As we shall see later, this may not always be done easily.

Visual shape or form provides another source of physiognomic (and synesthetic) meanings. In an early study, Krauss (1930) reported a high level of consensual agreement on the affective qualities of abstract line drawings—that lines connote such properties as gaiety, sadness, range, dawn, gold, and glass (see also Hevner, 1935). For example, three highly distinctive lines, which might be characterized as roughly approximating the visual representations of a square wave, a triangular wave, and a sine wave, were consistently matched with iron, silver, and gold, respectively. Perhaps best known of all, however, are Köhler's (1947) nonsense forms shown in Figure 1, the jagged *takete* and the lobular *maluma* (the latter originally called *baluma;* Köhler, 1929). According to Köhler (1947), "when asked to match the nonsense words 'takete' and 'maluma' with the two patterns . . . most people answer without hesitation" (p. 133).

The evocative powers of these two nonsense forms have been studied

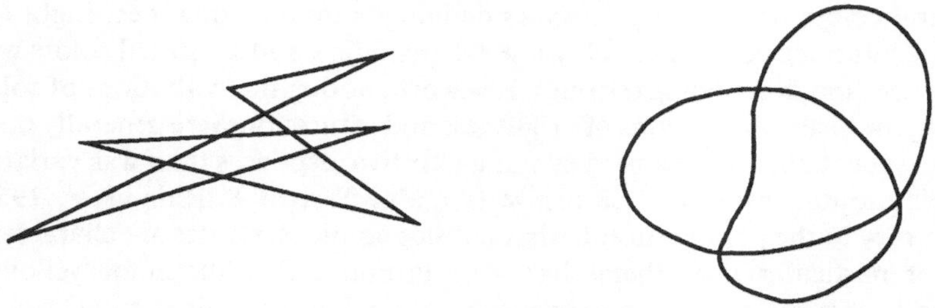

FIGURE 1 Köhler's (1929, 1947) nonsense figures *takete* (left) and *maluma* (right).

extensively by Lindauer (1990a, 1990b). In one study, Lindauer (1990b) asked participants to rate both forms, along with two additional, presumably physiognomically neutral forms (one a pattern of dots, the other described as ameba-like), on a variety of bipolar scales (e.g., peaceful–aggressive, relaxed–tense, quiet–noisy, cold–hot, bright–dark, big–small). As anticipated, *takete* was the most bright, rich in color, big, thin, aggressive, tense, strong, unfriendly, tough, noisy, fast, heavy, and sharp, whereas *maluma* was the most near, peaceful, relaxed, tender, friendly, quiet, and blunt. It was notable that even the "neutral" visual forms tended to produce some physiognomic responses, which is consistent with the notion that expressive characteristics are endemic and pervasive in perception. Subsequently, Lindauer (1990a) reported the presence of adaptation-like contrast effects. Using the same visual stimuli, he noted that ratings of the two neutral stimuli, especially ratings of their affective qualities, were modified when the forms followed either *maluma* or *takete,* for example, becoming more peaceful after the aggressive *takete* and more aggressive after the peaceful *maluma.*

Physiognomic properties presumably characterize responses to all visual forms including meaningful ones as well as arbitrary shapes (although the use of nonsense forms like *maluma* and *takete* undoubtedly helps disentangle potentially confounding meanings). Letters, words, and larger groupings of written material must be included among the meaningful forms whose shapes may evoke physiognomic reactions. Two studies of Koriat and Levy (1977, 1979) assessed "figural symbolism" in native speakers of Hebrew. The first study found clear correspondences in such dimensions as hardness and brightness between symbolic connotations of Japanese and Hindi vowels and the symbolic connotation of their orthographic representations; the second study showed reliable matching of Hebrew to Chinese antonyms.

Words have power of evocation acoustically, through their sound; visually, through their printed shape; and even proprioceptively, through the shape that the vocal apparatus takes in speaking them (to which Nabokov, 1949, attributed his own word–color synesthesia). The domain of "phonetic symbolism" is rife with examples in which meanings of words, either real or nonsense, can be conveyed physiognomically or cross-modally—through sound, orthography, or proprioception (Marks, 1975, 1978).

In Lindauer's (1991) view, as in Werner's (1957), physiognomic properties are primary in perception. Although physiognomic evocation is closely connected in many cases to synesthesia (a term that Lindauer uses for what I am calling cross-modal similarities), he infers from the predominance of affective, as compared with sensory, responses to words, drawings, and art that the initial reaction to a perceptual stimulus is affective or physiognomic, with synesthetic or cross-modal elaboration following suit within the course of perceptual processing. One should be cautious, however, about drawing inferences regarding temporal organization of perceptual processes on the basis of differences in response strength, especially from measures of the magnitude, frequency, or availability of verbal labels. That a measure of the strength of a response need not correlate with its priority in perceptual processing is suggested by the so-called Stroop phenomenon (e.g., Dyer, 1973). Under at least some circumstances, the speed and accuracy in making a perceptual discrimination, for example, between red and green, are greatly hindered by incongruent linguistic information (presenting the color red within the word *green* or the color green within the word *red*); yet speed and accuracy in making the complementary linguistic discrimination between *red* and *green* may be largely unaffected by incongruent perceptual information. This can happen even though the discrimination of hues precedes, and provides the ground for, the learning of color names. Indeed, considerations of this sort raise an important question about the role of overt or covert linguistic processes in processing nonlinguistic (perceptual) relations.

SYNESTHESIA AND CROSS-MODAL SIMILARITY: PERCEPTION AND LANGUAGE

Given that at least a handful of cross-modal similarities or equivalences is evident at birth or soon thereafter, one important consequence is that they precede the development of language. Presumably, therefore, the correspondence between the auditory qualities of pitch and the visual qualities of intensity, to use a prototypical example, is quintessentially perceptual in nature and not derived from or through language. Yet, it is also the case that synesthetic, cross-modal, or physiognomic meanings may reveal themselves through language as well as perception. Indeed, research into these phenomena as well as other kinds of perceptual metaphor rests largely on linguistic elaboration or verbal response. Spurred in large measure by the successes of semantic differentiation (Osgood, Suci, & Tannenbaum, 1957), a common method to assess meanings of perceptual stimuli is to ask a test population to rate each of them on verbally labeled rating scales. As Osgood et al. pointed out, their development of the semantic differential—a series of rating scales defined by bipolar adjectives—evolved from earlier work on the evaluation of synesthetic perception and tendencies.

Many of the very same correspondences evident in studies of cross-modal perception and similarity—for instance, the correspondences between auditory pitch and visual brightness and between loudness and brightness—reveal themselves when people evaluate words or phrases describing perceptual experiences. Not only do people match bright colors with loud and high-pitched tones and dim colors with soft and low-pitched tones, but people rate the word *sunlight* to connote greater loudness than the word *moonlight,* and rate the word *sneeze* to connote greater brightness than the word *cough* (Marks, 1982). In an unpublished experiment, I constructed a set of 25 words describing visual or auditory experiences, then asked 12 participants to rate the degree of similarity between the meanings of all possible pairs. Thus, on a given trial, both of the terms might have visual referents, both might have auditory referents, or one might have a visual referent and the other an auditory referent.

Multidimensional scaling revealed three basic dimensions underlying the judgments. The first two dimensions are represented in the left-hand panel of Figure 2, which plots each expression within a spatial array in which the distance between any pair of items reflects the overall degree of psychological dissimilarity. The first dimension distinguished weak from strong perceptual experiences, with such words as *soft, dim, murmur,* and *moonlight* located at the left-hand end of Dimension 1 and *loud, bright,* and *sunlight* at the right end. Dimension 1 also distinguished auditory experiences with regard to pitch, with *low sound* at the left end and *high sound* at the right. Pitch predominates, however, in Dimension 2, distinguishing words indicating low-pitched sounds such as *thunder, drum note,* and *cough* at one end and the higher pitched *whisper, squeak,* and *child's voice* at the opposite pole. A third, relatively minor, dimension distinguished all of the words referring to auditory experiences from those referring to visual experiences. This multimodal representation resembles only in very crude outline the result that I obtained when participants rated the degree of similarity of visual and auditory stimuli themselves (Marks, 1989), shown in the right panel of Figure 2. Judgments of tones and lights led to a much simpler representation of only two dimensions: one corresponding to the loudness of the sound, and the other, to the pitch of the sounds, with brightness running on a diagonal and thus aligned with both dimensions.

To be sure, the tendency for perceptual and verbal measures of cross-modal similarity to agree roughly is not surprising. If sounds and colors have particular physiognomic qualities that link them—better, that link the dimensions of pitch and loudness with lightness and brightness—then it is eminently reasonable to speculate that in learning relative labels such as *low pitch, loud, dim,* and *bright,* one comes to associate the attendant physiognomic properties that pertain to the corresponding percepts, much as Osgood et al. (1957) suggested. The correspondences between primary perceptual meanings and secondary linguistic ones need not be perfect—language and perception do not necessarily carve the world up in precisely the same way (cf. Miller & Johnson-Laird, 1976)—but the connections are nevertheless strong.

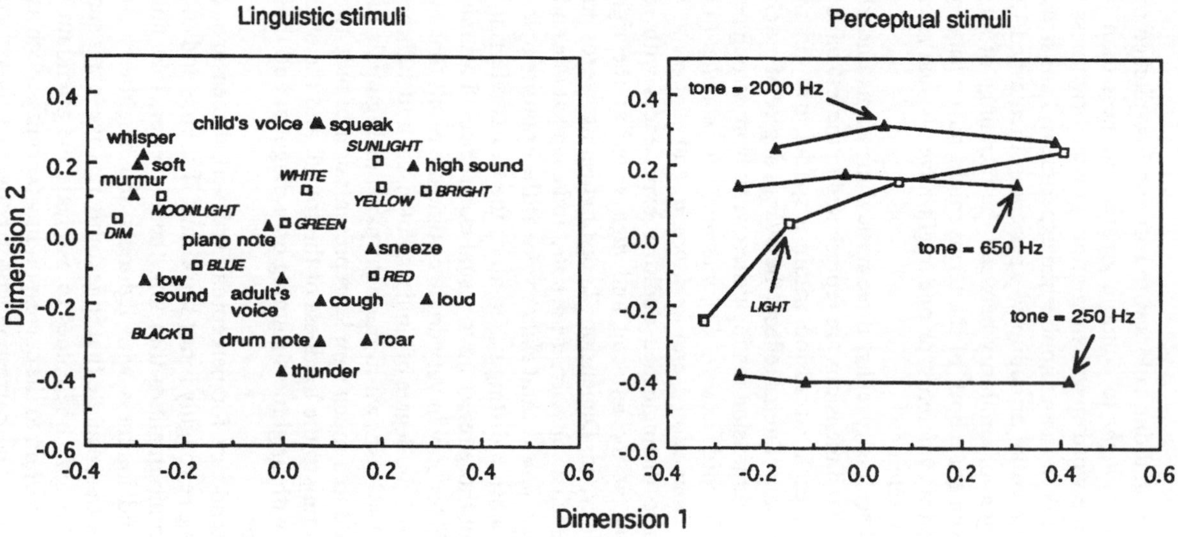

FIGURE 2 Multidimensional scaling derived from ratings of similarity between meanings of pairs of words describing visual and auditory experiences (left panel) compared to multidimensional scaling derived from ratings of similarity between pairs of visual and auditory stimuli (right panel).

SPEED AND ACCURACY IN FIGURATIVE PROCESSING

How might we evaluate perceptual metaphors that are instantiated in cross-modal, physiognomic similarities? One possible approach, popular with students of perceptual information processing, relies on measures of the speed and accuracy with which subjects can identify or classify stimuli. From such an information-processing perspective, response speed and accuracy provide two indexes of the ease, facility, or directness by which stimuli are encoded, decisions are made, and appropriate responses are selected. Everything else being the same, circumstances that impede perceptual processing will lead people to respond more slowly or less accurately, whereas circumstances that facilitate processing will lead people to respond relatively faster or more accurately. Consider the following paradigm: On each trial, a person sees one of two possible visual stimuli, for example, Köhler's (1929, 1947) *maluma* or *takete.* The task is to identify the stimulus as quickly as possible by pressing one of two keys, each being assigned to one of the stimuli. Of course, the person need not wait until the stimulus is fully processed before responding, and does not do so given the impetus to respond within a few hundred milliseconds. To identify the stimulus on a given trial, the person need process only a portion of the figure, perhaps a small rounded edge of *maluma* or a single acute angle of *takete.*

Let us assume, following suggestions made by Arnheim (1949), Werner (1957), and Lindauer (1991), that the figurative or physiognomic properties of forms like *maluma* and *takete* predominate in the earliest stages of perception and that intrinsic affective meanings characterize the first wave of the internal perceptual response. Bipolar in nature, the physiognomic response presumably consists of, or exemplifies, the properties associated with strength, aggression, and brightness in the case of *takete* (for convenience, call this the positive pole of the physiognomic dimension) or weakness, passivity, and darkness in the case of *maluma* (the negative pole). Assume further, following standard models of discriminative response (e.g., Grice, 1968; Link, 1992), that the person establishes two internal criteria or thresholds for pressing a key: one corresponding to the positive pole (respond to *takete*), and the other, the negative pole (respond to *maluma*). The participant presses a key once the physiognomic response (it may, of course, be any internal decisional variable) reaches an appropriate level corresponding to the value of the positive or negative criterion—this may well happen before the stimulus is consciously perceived geometrically as rounded or angular.

One of the propitious features of this scenario is that it leads to predictable consequences. Because physiognomic properties characterize the perceptual experiences aroused by stimulation of all the senses, it should be possible to find a pair of auditory stimuli that evoke covert physiognomic responses like those evoked by *maluma* and *takete.* Indeed, it may be that people rely on physiognomic properties of the spoken names "maluma" and "takete" when they match these names to the two forms—"maluma" being characterized by grave, which is to say low, frequencies and "takete," by acute or high frequencies. Low tones generate an initial physiognomic response resembling that evoked by *maluma,* whereas high tones generate a physiognomic response resembling that evoked by *takete.*

What will happen when the person is presented on each trial with a compound stimulus—one of the two visual forms plus one of the two sounds? Two of the bimodal combinations will be physiognomically congruent, and the other two will be incongruent. Given that the two physiognomic responses lie on different poles of a single dimension, it follows that presenting a low tone along with *maluma* reinforces the physiognomic response to the visual stimulus, increasing the degree of negativity as defined previously, and presenting a high tone with *takete* likewise reinforces the degree of positivity. Reversing the cross-modal pairings, however, by presenting a low tone with *takete* or a high tone with *maluma,* dilutes the physiognomic response evoked by the visual stimulus alone, shifting the overall level of the response away from the positive or negative pole and back toward neutral. Assuming, first, that the overt response—the key press—depends on the overall level of the covert, physiognomic response, and second, as seems reasonable, that there is variability associated with the responses to every stimulus, then it follows that presenting a low-frequency or a high-frequency auditory stimulus together with the visual stimulus will impede or facilitate the behavior. Overt responses to physiognomically congruent stimuli should be faster or more accurate than responses to physiognomically incongruent stimuli.

This is precisely the result obtained by Marks (1987, Experiment 4). Participants pressed one of two keys in order to identify a visual stimulus as either (a) a rounded, U-shape; or (b) an angular, V-shape. At the same time, the participants heard a complex acoustic signal with a relatively low or high fundamental frequency that was randomly associated with the visual shape and thus wholly uninformative. On trials in which the participants saw the rounded form (an ersatz *maluma*), responses were faster and more accurate if the tone's frequency was low rather than high, and on those trials in which the participants saw the acute form (an ersatz *takete*), responses were faster and more accurate if the tone's frequency was high rather than low. That is, performance was superior whenever the irrelevant sound was physiognomically similar to the visual stimulus being identified relative to performance when the irrelevant sound was physiognomically dissimilar to the visual stimulus. It was not possible, however, to determine whether the interactions actually represented both facilitation of processing in the case of congruence and inhibition in the case of incongruence or whether they represented nothing but inhibition.

Visual and auditory stimuli bearing no literal connection or physical similarity can nevertheless interact, presumably on the basis of their perceptually metaphorical bond (if the stimuli are connected to the criterion). Furthermore, the classification task requires no verbalization by the participant and, hence, no overt linguistic evaluation. Moreover, these interactions reveal themselves in responses typically completed within approximately 0.3 sec from the onset of the stimuli, a good portion of which (100 msec or so) is undoubtedly devoted to revving up the motor system and executing the finger press. So we may conservatively infer that elementary perceptual stimuli presented to different sense modalities can effectively evoke physiognomic responses, which in turn combine and initiate a decision to respond motorically, within 200 msec from their onset and perhaps even more quickly.

Similar findings have been obtained in analogous tasks using a variety of cross-modal pairings. Tones differing in pitch affect the identification of visual stimuli differing in brightness, and lights differing in brightness affect identification of tones differing in pitch (Marks, 1987; Melara, 1989). Loudness affects the identification of brightness, and brightness affects loudness (Marks, 1987). Pitch affects the identification of vertical visual position, and vertical visual position affects pitch (Ben-Artzi & Marks, 1995; Bernstein & Edelstein, 1971; Melara & O'Brien, 1987). The significance of these particular pairings of perceptual dimensions is clear. Stimuli that interact in tasks of speeded identification are the very ones that people judge to be cross-modally similar; so it is likely that measures of both cross-modal similarity and differences in efficiency of processing reflect communalities in underlying physiognomic responses. Visual and auditory stimuli that evoke the same or different physiognomic responses are judged as more or less similar and are processed more or less quickly and accurately. Crucial in this respect is the finding that interactions are absent when one pairs those visual and auditory stimuli, such as dark- and light-colored surfaces with soft and loud sounds, that lack cross-modal or synesthetic similarity (Marks, 1987).

Tasks requiring participants to identify stimuli rapidly and accurately have the virtue of being exquisitely sensitive to factors that may impede or facilitate processing, including the processing of perceptual metaphors. As I already indicated, one of the benefits of these procedures is that they require no overt verbal evaluation of the stimuli. At the same time, this is a severe limitation because tasks of stimulus identification provide little insight into how people actually understand, interpret, or assign meanings. Even so, the use of rapid, nonverbal responses, which people can make as though they operate on "automatic pilot," may help to minimize the role of linguistic processes, a particular virtue if one's goal is to decompose the stages involved in processing perceptual metaphors.

A ROLE FOR LINGUISTIC PROCESSES?

Having said this, we should not be too quick to assume that even the rapid processing of nonlinguistic, perceptual stimuli—like the processing I assumed to involve implicit physiognomic responses—wholly evades the engagement of verbal mechanisms. As a matter of fact, there are good reasons to consider seriously the possibility that linguistic processes may impinge on perceptual ones (perhaps as perception or imagery may impinge on processing of figurative language, e.g., Honeck & Kibler, 1984). Consider the series of speeded-identification experiments reported by Walker and Smith (1984). On each trial, participants pressed one of two keys to classify a visual stimulus, and this visual stimulus was always accompanied by an irrelevant sound that could take on either low- or high-sound frequency. So far, the experiment resembles those just described. In Walker and Smith's study, however, all of the visual stimuli were words comprising pairs of antonymic adjectives such as *low–high, sad–happy, big–little, weak–strong, slow–fast,* and *passive–active.* Within a given block of trials, participants would see four possible words (e.g., *sharp, blunt, bright,* or *dull*) randomly presented one at a time; the participant was to categorize the stimulus as, in this example, "sharp" or "bright" (by depressing one key) or as "blunt" or "dull" (by depressing another key). Thus, the paradigm parallels that of Marks (1987, Experiment 4), except that Walker and Smith's participants responded to adjectives (words), whereas Marks's participants responded to rounded and angular visual shapes.

In Walker and Smith's (1984) experiment, the sound frequency of the tone interacted with the word: The participants responses were relatively quick when the word's meaning and tone matched cross-modally or physiognomically (e.g., the word *dull* or *blunt* accompanied by the low-frequency tone or the word *sharp* or *bright* accompanied by the high-frequency tone) but were relatively slow when the word and tone did not match (e.g., *sharp* or *bright* accompanied by the low tone or *dull* or *blunt* accompanied by the high tone). Thus, covert responses to visually presented words, responses activated through semantic mechanisms, can interact with covert physiognomic responses evoked perceptually by tones. Melara and Marks (1990) and Ben-Artzi and Marks (1995) reported similar findings (see also Kaden, Wapner, & Werner, 1955).

Just because linguistic and nonlinguistic stimuli interact much as two nonlinguistic stimuli do does not mean that the mechanisms underlying the two kinds of interaction are identical. Where interactions between linguistic and nonlinguistic stimulus dimensions, such as visual words and tones varying in pitch, undoubtedly involve language-based semantic processes, it is perfectly possible that interactions between two nonlinguistic stimulus dimensions, such as visual stimuli differing in lightness and tones differing in pitch, occur without verbal mediation.

At the very least, however, observations with linguistic stimuli raise the possibility that implicit semantic processes may mediate, in part or in whole, the cross-modal interactions that I attributed to common physiognomic response. According to this alternate view, intersensory interactions, regardless of whether any of the stimuli are linguistic, take place only after all of the stimulus information is encoded or recoded into a common semantic format. Such a hypothesis contradicts the notion that interactions rely entirely on undifferentiated physiognomic responses that unfold early in perceptual processing. It is also hard to reconcile with models (e.g., Smith & Kemler Nelson, 1978) that assume that cross-dimensional interactions in tasks of speeded stimulus identification result from early holistic processing. More important, it suggests a possible role of linguistic mechanisms in the processing of perceptual metaphors.

Are cross-modal interactions mediated semantically? Although only a few studies have sought to answer this question, and not all of the evidence is in, several observations (summarized in Marks, 1995) imply that linguistic mechanisms do intrude into the processing of perceptual attributes of stimuli. Consider an experiment in which participants try to identify the color of a visual stimulus as dark or light with each visual stimulus accompanied by a tone that may be low or high in frequency. Instead of constituting a simple patch of color, however, the the visual stimulus could be a word, for example, *black,* printed in dark or light color. The word is constant throughout a given experimental session (with some participants, the word is always *white*) and, thus, should have no bearing on the response.

Recall the two possible accounts just suggested. According to one, every stimulus first evokes a physiognomic response, which precedes delineation of such characteristics as its precise shape, color, or (presumably) the word itself. Furthermore, physiognomic processing precedes semantic processing. Consequently, the fact that the visual stimulus is always a word—even one that is semantically related to the physiognomic response to its color—should be irrelevant. The organismic, physiognomic response to the color component, as Werner (1957) might have said, should interact with the corresponding physiognomic response to the accompanying sound before the visual stimulus is elaborated linguistically. Thus, a purely perceptual, physiognomic account predicts a strong interaction between lightness and pitch regardless of whether the visual stimulus is a color word or a nonlinguistic patch of color.

A semantic account makes a different prediction. In this account, all of the processing involves semantic codes. Congruent combinations of color and sound (dark with low pitch and light with high pitch) are processed more efficiently than incongruent combinations (dark with high pitch and light with low pitch) because the semantic codes of the visual and auditory stimuli share more or fewer common features, respectively. Given the potency of linguistic stimuli in evoking semantic responses, presentation of a constant word should invite a constant implicit semantic response, which, in turn, should act to dampen the differential semantic effects produced by dark and light colors themselves and, thus, to the same extent, should dampen the interaction between color and tone. Basically, presenting a constant word should serve to suppress the participant's tendency to implicitly label the colors semantically.

Are interactions in response speed and accuracy smaller when the visual stimulus comprises a color within a constant word rather than a nonlinguistic patch of color? They are. On the whole, interactions in response time dropped by 40% (overall response times with congruent colors and tones were 13.6 msec smaller than those with incongruent ones when participants saw light- and dark-colored patches but only 8.1 msec smaller when participants saw light- and dark-colored words), and the interactions in accuracy wholly disappeared (errors were 4% fewer with congruent compared to incongruent stimuli in the first case but 0.3% greater in the second). That is, the congruence of the tone mattered much more when the visual stimulus was a patch of color than when it was a constant word. This outcome suggests a central role for semantic codes in intersensory interactions; semantic processes could account in part, and maybe in whole, for the cross talk between colors and tones.

The findings just described imply that semantic processes play some role in the interactions between physiognomically similar visual and auditory stimuli. Do they play the whole role? Or is there any evidence that purely perceptual processing also has a part? To seek clear evidence of a physiognomic, perceptual process, I measured cross-modal interactions in a paradigm that requires the participants to attend to both perceptual and linguistic stimulus dimensions and, by doing so, that encourages semantic coding. Are there interactions based in nonlinguistic processing when linguistic processes are emphasized?

Again, I used as visual stimuli two color words (*black* or *white*), each of which could differ in color (dark and light). Each combination of word and color was accompanied by a tone taking on one of two frequencies (low or high). Once more, the auditory stimulus was irrelevant to the participant's task, which in this case was to decide whether, on each trial, the perceptual and linguistic components agreed or disagreed—that is, whether the stimulus was visually congruent or visually incongruent. Participants pressed one key if the visual stimulus consisted of the word *black* printed in dark color or the word *white* printed in light color (visually congruent stimulus) and another key if the visual stimulus consisted of either *black* printed in light color or *white* printed in dark color (visually incongruent stimulus). Participants cannot respond until they assess both the color and the word and then compare them; therefore, the paradigm relies heavily on linguistic processing because it is likely that participants must explicitly recode the color into its name, which then can be compared with the color word. Evidence of purely perceptual interactions surviving all of the semantic processing imposed by this task would suggest a role for primary, physiognomic responses in mediating the cross-modal interaction.

In this divided-attention paradigm, the visually congruent stimuli are of little interest; because the values of the perceptual and semantic components of the visual stimulus are positively correlated, the visual stimuli should interact with the auditory stimulus in the same manner regardless of whether the perceptual color is recoded semantically. The visually incongruent stimuli, however, provide a crucial test because the values of the perceptual and semantic components are negatively related. If sounds and colors interact physiognomically, before the visual stimulus is processed as a word, then the tones should interact more with the colors than with the color words. Processing should be relatively efficient—relatively rapid and accurate—when the participants receive a low-frequency tone with a black color, even though the word is *white,* and when they receive a high-frequency tone with a light color, even though the word is *black.* Processing should be relatively inefficient when the participants receive a low tone with a light color, even though the word is *black,* and when they receive a high tone with a black color, even though the word is *white.* A linguistic model makes the opposite prediction: Tones should interact more with the color words than with the colors themselves.

The results appear in Figure 3. To display the data efficiently, for each of the eight possible stimuli, the figure plots the percentage of errors against the response time. The four data points lying in the upper left portion of the figure show the results obtained with the visually congruent, (nondiagonostic) stimuli. Processing is good (fewest errors and smallest response times) when the pitch of the tone is congruent with both the color and the color word (word *black* in dark color with low pitch, word *white* in light color with high pitch) and poorer when pitch is incongruent with both (*black* in dark color with high pitch, *white* in light color with low pitch). This is as expected. The visually incongruent stimuli (right side of the figure), however, show a very different pattern. Errors, mostly, are affected, and the thrust of the interaction implicates the pitch of the tone and the color word rather than the color. Thus, performance is particularly good when the participants had to identify the word *black* presented in light color and the pitch was low or when they had to identify the word *white* presented in dark color and the pitch was high; in both cases, pitch is congruent with the label—with the word—but is incongruent with the color in which the word was printed. Finally, the overall difference between responses to visually congruent stimuli (fast RTs, many errors) and to visually incongruent stimuli (slow RTs, few errors) may itself represent a tradeoff between speed and accuracy and, therefore, may repre-

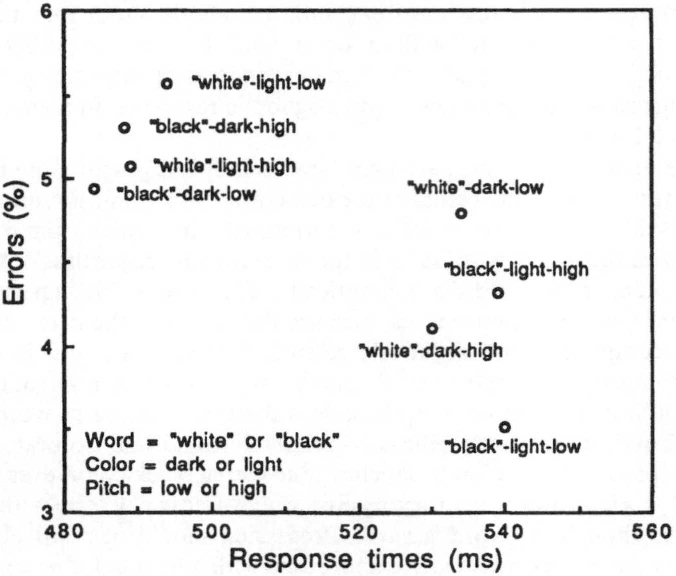

FIGURE 3 Errors plotted against response times for each of eight stimuli, classified as either visually congruent (word *black* printed in dark color or word *white* printed in light color or visually incongruent (*black* in light color or *white* in dark color). Each combination of word and color was accompanied by an irrelevant tone, which could be either low or high frequency.

sent an asymmetry between the criteria used for the two responses.

The results of this experiment suggest that semantic processes play a prime role in cross-modal interaction. However, it is important to bear in mind that this paradigm of divided attention is especially conducive to semantic processing. The failure to find clear evidence of a perceptual, physiognomic interaction between pitch and color need not mean that physiognomic interactions did not exist, only that the paradigm failed to reveal them. Moreover, the pattern of behavior may also depend on the particular response system being tapped (e.g., Hughes, Reuter-Lorenz, Nozawa, & Fendrich, 1994). On the other hand, the results do reinforce the importance of considering when and how linguistic encoding and recoding affect perceptual processing and when and how language-based processes intrude into responses made to purely perceptual stimulus dimensions. Such considerations lead to the question of whether semantic recoding might be involved in other measures of cross-sensory figuration and physiognomic perception. Studies in which subjects are asked to evaluate the meanings of perceptual stimuli, even studies in which participants are asked to match perceptual stimuli across different modalities, could rely on implicit semantic responses.

Whether, or to what extent, the crucial processes are physiognomically perceptual or semantic remains unclear. What is clearer is the need to develop more critical research paradigms, whose results can be interpreted within explicitly articulated theories and models of behavior. Research on figurative processing in perception would benefit from closer contact with mainstream research in cognitive psychology. In this regard, it is notable that Kornblum, Hasbroucq, and Osman (1990) have incorporated cross-modal associations into their model of stimulus-response compatibility.

ENDINGS AND BEGINNINGS

Even if some perceptual metaphors end up being mediated linguistically, their origins appear to be wholly in perception itself, starting within perceptual processes before being overlaid and dominated by linguistic ones. A recent study by Braaten (1993) reported evidence of interactions in 6-month-old infants between visual stimuli varying in spatial location and congruent and incongruent auditory stimuli varying in sound frequency. Braaten's paradigm, which used response time for viewing as a measure, is formally analogous to the speeded identification procedure used in adults, in which interactions appear to be mediated at least partly by language. Given that interactions in infancy are purely perceptual, it is only later in development that language begins to intrude.

TAKING QUALITATIVE EXPERIENCES SERIOUSLY

There is another way to interpret all of the findings just described, and it bears directly on the issue of how to characterize cross-modal metaphors. Perhaps, all expressive meanings—whether generated perceptually (i.e., physiognomically) or semantically—are processed through a single, common amodal mechanism. Cross-modal similarities and interactions could reflect the concurrent activation of "features" that are themselves neither auditory nor visual, neither strictly perceptual nor strictly linguistic (an analogue can be found in Honeck & Temple's, 1994, conceptual base theory, in which abstract representations underlie the figurative meanings of proverbs that can apply to pictorial as well as verbal examples). A piece of strong evidence favoring this view comes from the fact that lightness and pitch provide a prototype of a metaphorical relation between vision and hearing, yet the attributes of these dimensions lack common semantic labels. To infer that pitch and lightness interact at a semantic level, given that the labels *low* and *high* differ from the labels *dark* and *light,* implies that the interactions themselves rely on a highly abstract code. From this, it is a small step to postulating that all varieties of cross-modal similarity and interaction rely on abstract codes—that perceptual and linguistic stimuli alike activate abstract features, or "semes," which are intrinsically neither perceptual nor linguistic.

Is there a link between the amodal processing of perceptual or semantic attributes and other kinds of amodal processing? If, following Gibson (e.g., 1979), the perception of a chair includes the functional affordance it provides of "sittingness," then, following Werner (1957), does not the perception of a dark cloud include the physiognomic affordance of its "threateningness"? Researchers in the Gibsonian tradition (e.g., Verbrugge & McCarrell, 1977) suggest that metaphors characterize resemblances between things in the world. Working in this "direct realist" vein, Dent-Read and Szokolszky (1993), who recognize that the notion of metaphor applies to pictures and to actions as well as language, distinguished metaphor from other activities on the basis of two criteria: First, metaphor "involves the detection and use of structural or dynamic properties that remain invariant across kinds," and second, "one kind . . . is perceived in terms of another kind" (p. 230). An example they mention is calling small, isolated, active DNA segments "islands," a metaphor based, in their view, on perceiving the resemblance that exists between the invariant structural characteristics of nucleic acids and isolated bodies of land.

If we adapt an ecological account of metaphors to physiognomic resemblances, we could infer that the communality inherent in dynamic, physiognomic properties—in what makes a dark cloud threatening or a bird's song joyful—reflects their resemblances in the world. Dark clouds are like threats, particular melodies like physical expressions of happiness—much as Plato said that the prototype of "triangularity" resembles all triangles, actual and potential.

This interpretation is reminiscent of Arnheim's (1949) theory of expressiveness, based in the Gestalt notion of "isomorphism, according to which processes which take place in different media may be nevertheless similar in their structural organization" (p. 157). In following the tradition of Gestalt psychology, however, Arnheim's interpretation of physiognomic perception remains more firmly psycho-physical—more willing to place metaphor within a context concerned with the phenomenal properties of experience—than are accounts hammered in a Gibsonian forge. I must confess greater sympathy with Arnheim's approach. Although I recognize that many metaphors, whether in words, photographs, music, or dance, are grounded in physical resemblances, nevertheless, other figurative presentations (in particular, ones expressing cross-modal, physiognomic, affective properties) reflect common meanings that are better considered in mentalistic terms. It is the experience of a certain color that makes a necktie appear loud and the experience of certain sunrises that made Kipling describe them as "coming up like thunder" (even if he never saw China 'cross the bay). Kennedy (1982) came to a related conclusion when he noted that the very notion of metaphor implies intentionality in its deviation from some norm.

Although physical resemblances might underlie phenomenal similarities—very likely there is something in common in the neural activity underlying states of joy and certain kinds of music, although it need not be Gestalt-like isomorphism—this is not what metaphors that invoke phenomenal states are all about. They are about the phenomenal experiences themselves.

Many psychologists continue to tread carefully around such issues, for there are few, if any, good empirical or conceptual tools with which to grip the problems associated with qualitative aspects of experience. Although it may be tempting to attribute this unfortunate state of affairs to the rejection of introspectionism, it is not at all clear that a return to introspection per se would fix our hold. The language of science is essentially one that makes distinctions along physical or pseudo-physical dimensions. Nor does philosophy fare much better, even when it is recognized that *qualia* may matter. A recent attempt at a philosophical account of sensory qualities (Clark, 1993) essentially relies on psychophysical analysis through measures of discriminability and multidimensional scaling—once more, reducing qualitative variation to "quality-less," quantitative difference.

SIGNS AND SYMBOLS: NATURAL AND CONVENTIONAL

If metaphor is a kind of symbolic activity, then "symbolic activity" needs to include perception as well as language. My focus on various affective or physiognomic properties of perceptual stimuli reflects my conviction that there are natural correspondences between experiences in different sense modalities, and that these seem to be nothing less than "hard wired." In other words, cross-modal similarities may form a subset of "natural signs" or "natural symbols." That certain signs or symbols may be natural or unconditional and others conventional or conditional is a recurrent notion going back at least to Plato's *Cratylus.*

Stafford (1979) recounted and analyzed the early 19th-century attempts by Humbert de Superville to use physiognomic characteristics to establish a natural order of signs in the arts. Properties of color and line in Humbert's view provide elements of a universal grammar for expressing emotional, intellectual, and even moral attributes—for example, horizontal lines are linked to notions of equilibrium, order, brightness, and whiteness, whereas expanding lines are linked to agitation, scattering, explosiveness, and redness. L.-B. Castel (1725), inventor of what was probably the first color organ, followed Newton in believing in a natural connection between spectral colors and notes of the diatonic musical scale, and various artists, from Kandinsky to Messiaen, have engaged in, or flirted with, analogous schemes.

Certainly, no system of natural signs can account for the entire gamut of metaphors. By focusing on common sensory and physiognomic processes, I have left untapped any consideration of the ways that producing and understanding metaphors can invoke processes of abstraction and selection or, more generally, constructive processes. Experience, perception, and language alike are manifold and multidimensional. Metaphors are, or can be, creative.

In their study of the symbolic meanings of visual forms—a square and a circle—Liu and Kennedy (1993) found considerable consensual agreement but argued that "the symbolic relations between the shapes and the words . . . are mostly new to our subjects . . . [who,] we suggest, are constructing an answer, rather than simply retrieving one" (p. 453). In a related vein, several investigators (e.g., Glucksberg & Keysar, 1990; Medin, Goldstone, & Gentner, 1993) questioned the usefulness of explaining metaphoricity in terms of similarity, noting, for example, that what are needed are explanations of how people decide which characteristics make items similar. Medin et al. argue that "similarity involves far more than a simple computation over a set of fixed features. Instead, similarity is always dynamic, is often inherently asymmetrical, and discovers and aligns features rather than just adding them up" (p. 275). Again, a central property of recognizing metaphor as metaphor comes from acknowledging its intentionality (Kennedy, 1982).

In any case, a sharp distinction between natural and conventional signs is inadequate. Signs or symbols may be more or less natural. Many cultures have developed elaborate symbologies and mythologies centering on perception—not just in vision and hearing but also in taste, touch, and smell (in this regard, see the recent edited volume on the cultural anthropologies of the senses; Howes, 1991), some of which contain elements of natural as well as conventional symbol systems. Speed lines are conventional, their role not being recognized immediately by the uninitiated; as Kennedy (1993) noted, people work them out, the blind doing this much as sighted people do. Arnheim (1969) suggested that

> all media of representation can rely on isomorphic and on non-isomorphic references. They are partly analogues, partly signs. In principle, there is no difference in this respect between verbal and non-verbal languages. The most important difference in practice is one of ratio. In the visual arts or in music, for example, strictly non-isomorphic references are exceedingly rare. In verbal language, they do most of the work. (p. 251)

It is perhaps because artistic media so often rely on isomorphic representations that figurative, metaphorical, and physiognomic properties become salient in works of art—to color and line in painting or to orthography and phonetic attributes of words in reading or listening to poetry (see Marks, 1978). However, the question remains: To what extent do we transform what we see and hear into another set of signs, the convention-based words with which we talk about them, both to others and to ourselves?

ACKNOWLEDGMENTS

I thank Joseph Stevens and John Kennedy for many thoughtful comments on earlier versions of this article. Preparation of this article was supported by Research Grant 5R01DC 00271–12 from the National Institute on Deafness and Other Communication Disorders, National Institutes of Health.

REFERENCES

Arnheim, R. (1949). The Gestalt theory of expression. *Psychological Review, 56,* 156–171.

Arnheim, R. (1969). *Visual thinking.* Berkeley: University of California Press.

Bar-Cohen, S., Harrison, J., Goldstein, L. H., & Wyke, M. (1993). Coloured speech perception: Is synaesthesia what happens when modularity breaks down? *Perception, 22,* 419–426.

Ben-Artzi, E., & Marks, L. E. (1995). Visual-auditory interaction in speeded identification: Role of stimulus difference. *Perception & Psychophysics, 57,* 1151–1162.

Bernstein, I. H., & Edelstein, B. A. (1971). Effects of some variations in auditory input upon visual choice reaction time. *Journal of Experimental Psychology, 87,* 241–247.

Black, M. (1962). *Models and metaphors: Studies in language and philosophy.* Ithaca, NY: Cornell University Press.

Börnstein, W. (1936). On the functional relations of the sense organs to one another and to the organism. *Journal of General Psychology, 15,* 117–131.

Bowers, J. W. (1990). Dating "a figure of thought." *Metaphor and Symbolic Activity, 5,* 249–250.

Braaten, R. (1993, November). *Synesthetic correspondence between visual location and auditory pitch in infants.* Paper presented at the annual meeting of the Psychonomic Society, Washington, DC.

Castel, L.-B. (1725). Clavecin par les yeux, avec l'art de peindre les sons, & toutes sortes de pièces de musique [Harpsichord by sight, with the art of painting sounds, and all kinds of musical works]. *Mercure de France,* 2552–2577.

Clark, A. (1993). *Sensory qualities.* Oxford, England: Clarendon.

Cytowic, R. E. (1989). *Synesthesia: A union of the senses.* New York: Springer-Verlag.

Damhorst, M. L., & Reed, A. P. (1986). Clothing color value and facial expression: Effects on evaluations of female job applicants. *Social Behavior and Personality, 14,* 89–98.

D'Andrade, R., & Egan, M. (1974). The colors of emotions. *American Ethnologist, 1,* 49–63.

Dent-Read, C. H., & Szokolszky, A. (1993). Where do metaphors come from? *Metaphor and Symbolic Activity, 8,* 227–242.

Dyer, F. N. (1973). The Stroop phenomenon and its use in the study of perceptual, cognitive, and response processes. *Memory & Cognition, 1,* 106–120.

Feré, C. (1892). *La pathologie des emotions* [The pathology of the emotions]. Paris: Alcan.

Frank, M. G., & Gilovich, T. (1988). The dark side of self- and social perception: Black uniforms and aggression in professional sports. *Journal of Personality and Social Psychology, 54,* 74–85.

Gibbs, R. W. (1984). Literal meaning and psychological theory. *Cognitive Science, 8,* 275–304.

Gibson, J. J. (1979). *The ecological approach to visual perception.* Boston: Houghton-Mifflin.

Glucksberg, S., & Keysar, B. (1990). Understanding metaphorical comparisons: Beyond similarity. *Psychological Review, 97,* 3–18.

Grice, G. R. (1968). Stimulus intensity and response evocation. *Psychological Review, 75,* 359–373.

Haskell, R. E. (1987). Giambattista Vico and the discovery of metaphoric cognition. In R. E. Haskell (Ed.), *Cognition and symbolic structures: The psychology of metaphoric transformation* (pp. 67–82). Norwood, NJ: Ablex.

Hevner, K. (1935). Experimental studies of the affective value of colors and lines. *Journal of Applied Psychology, 19,* 385–398.

Honeck, R. P., & Kibler, C. (1984). The role of imagery, analogy, and instantiation in proverb comprehension. *Journal of Psycholinguistic Research, 13,* 393–414.

Honeck, R. P., & Temple, J. G. (1994). Proverbs: The extended conceptual base and great chain metaphor theories. *Metaphor and Symbolic Activity, 9,* 85–112.

Howes, D. (1991). *The varieties of sensory experience: A sourcebook in the anthropology of the senses.* Toronto, Canada: University of Toronto Press.

Hughes, H. C., Reuter-Lorenz, P. A., Nozawa, G., & Fendrich, R. (1994). Visual-auditory interactions in sensorimotor processing: Saccades versus manual responses. *Journal of Experimental Psychology: Human Perception and Performance, 20,* 131–153.

James, W. (1890). *The principles of psychology.* New York: Holt.

Kaden, S. E., Wapner, S., & Werner, H. (1955). Studies in physiognomic perception: II. Effect of directional dynamics of pictured objects and of words on the position of the apparent horizon. *Journal of Psychology, 39,* 61–70.

Karwoski, T. F., & Odbert, H. S. (1938). Color-music. *Psychological Monographs, 50*(2, Serial No. 222).

Karwoski, T. F., Odbert, H. S., & Osgood, C. E. (1942). Studies in synesthetic thinking: II. The role of form in visual responses to music. *Journal of General Psychology, 26,* 199–222.

Kennedy, J. M. (1982). Metaphor in pictures. *Perception, 11,* 589–605.

Kennedy, J. M. (1990). Metaphor—Its intellectual basis. *Metaphor and Symbolic Activity, 5,* 115–123.

Kennedy, J. M. (1993). *Drawing & the blind: Pictures to touch.* New Haven, CT: Yale University Press.

Kennedy, J. M., Green, C. D., & Vervaeke, J. (1993). Metaphoric thought and devices in pictures. *Metaphor and Symbolic Activity, 8,* 243–255.

Köhler, W. (1929). *Gestalt psychology.* New York: Liveright.

Köhler, W. (1947). *Gestalt psychology* (2nd ed.). New York: Liveright.

Koriat, A., & Levy, I. (1977). The symbolic implications of vowels and of their orthographic representations in two natural languages. *Journal of Psycholinguistic Research, 6,* 93–103.

Koriat, A., & Levy, I. (1979). Figural symbolism in Chinese ideographs. *Journal of Psycholinguistic Research, 8,* 353–365.

Kornblum, S., Hasbroucq, T., & Osman, A. (1990). Dimensional overlap: Cognitive basis for stimulus-response compatibility—A model and taxonomy. *Psychological Review, 97,* 253–270.

Krauss, R. (1930). Über graphischen Ausdruck [On graphic expression]. *Zeitschrift für Angewandte Psychologie, 48,* 1–141.

Lakoff, G., & Johnson, M. (1980). *Metaphors we live by.* Chicago: University of Chicago Press.

Lawson, K. R., & Turkewitz, G. (1980). Intersensory function in newborns: Effect of sound on visual preferences. *Child Development, 51,* 1295–1298.

Lewkowicz, D., & Turkewitz, G. (1980). Cross-modal equivalence in early infancy: Auditory-visual intensity matching. *Developmental Psychology, 16,* 597–607.

Lindauer, M. (1990a). The effects of the physiognomic stimuli *taketa* and *maluma* on the meanings of neutral stimuli. *Bulletin of the Psychonomic Society, 28,* 151–154.

Lindauer, M. (1990b). The meanings of the physiognomic stimuli *taketa and maluma. Bulletin of the Psychonomic Society, 28,* 47–50.

Lindauer, M. (1991). Physiognomy and verbal synesthesia compared: Affective and intersensory descriptors of nouns with drawings and art. *Metaphor and Symbolic Activity, 6,* 183–202.

Link, S. (1992). *The wave theory of similarity and difference.* Hillsdale, NJ: Lawrence Erlbaum Associates, Inc.

Liu, C. H., & Kennedy, J. M. (1993). Symbolic forms and cognition. *Psyke & Logos, 14,* 441–456.

Marks, L. E. (1975). On colored-hearing synesthesia: Cross-modal translations of sensory dimensions. *Psychological Bulletin, 82,* 303–331.

Marks, L. E. (1978). *The unity of the senses: Interrelations among the modalities.* New York: Academic.

Marks, L. E. (1982). Bright sneezes and dark coughs, loud sunlight and soft moonlight. *Journal of Experimental Psychology: Human Perception and Performance, 8,* 177–193.

Marks, L. E. (1987). Auditory-visual interactions in speeded discrimination. *Journal of Experimental Psychology: Human Perception and Performance, 13,* 384–394.

Marks, L. E. (1989). On cross-modal similarity: The perceptual structure of pitch, loudness, and brightness. *Journal of Experimental Psychology: Human Perception and Performance, 15,* 586–602.

Marks, L. E. (1995). Intermodal similarity and cross-modality matching. In R. D. Luce, M. D'Zmura, D. Hoffman, G. J. Iverson, & A. K. Romney (Eds.), *Geometric representations of perceptual phenomena: Papers in honor of Tarow Indow's 70th birthday* (pp. 207–233). Mahwah, NJ: Lawrence Erlbaum Associates, Inc.

Marks, L. E., Hammeal, R. J., & Bornstein, M. H. (1987). Perceiving similarity and comprehending metaphor. *Monographs of the Society for Research in Child Development, 52*(1, Serial No. 215).

Maurer, D. (1993). Neonatal synesthesia: Implications for the processing of speech and faces. In B. de Boysson-Bardies, S. de Schoenen, P. Jusczyk, P. McNeilage, & J. Morton (Eds.), *Developmental neurocognition: Speech and face processing in the first year of life* (pp. 109–124). Dordrecht, The Netherlands: Kluwer.

Maurer, D., & Maurer, C. (1988). *The world of the newborn.* New York: Basic Books.

Medin, D. L., Goldstone, R. L., & Gentner, D. (1993). Respects for similarity. *Psychological Review, 100,* 254–278.

Melara, R. D. (1989). Dimensional interaction between color and pitch. *Journal of Experimental Psychology: Human Perception and Performance, 15,* 69–79.

Melara, R. D., & Marks, L. E. (1990). Processes underlying dimensional interactions: Correspondences between linguistic and nonlinguistic dimensions. *Memory & Cognition, 18,* 477–495.

Melara, R. D., & O'Brien, T. P. (1987). Interaction between synesthetically corresponding dimensions. *Journal of Experimental Psychology: General, 116,* 323–336.

Miller, G. A., & Johnson-Laird, P. N. (1976). *Language and perception.* Cambridge, MA: Belknap.

Nabokov, V. (1949, April 9). Portrait of my mother. *New Yorker, 27,* 33–37.

Odbert, H. S., Karwoski, T. F., & Eckerson, A. B. (1942). Studies in synesthetic thinking: I. Musical and verbal association of color and mood. *Journal of General Psychology, 26,* 153–173.

Osgood, C. E., Suci, G. J., & Tannenbaum, P. H. (1957). *The measurement of meaning.* Urbana: University of Illinois Press.

Richards, I. A. (1936). *The philosophy of rhetoric.* Oxford, England: Oxford University Press.

Riggs, L. A., & Karwoski, T. F. (1934). Synaesthesia. *British Journal of Psychology, 25,* 29–41.

Roffler, S. K., & Butler, R. A. (1968). Localization of tonal stimuli in the vertical plane. *Journal of the Acoustical Society of America, 43,* 1260–1266.

Schlessinger, L. B. (1980). Physiognomic perception: Empirical and theoretical perspectives. *Genetic Psychology Monographs, 101,* 71–97.

Smith, L. B., & Kemler Nelson, D. G. (1978). Overall similarity in adults' classification: The child in all of us. *Journal of Experimental Psychology: General, 113,* 137–159.

Stafford, B. M. (1979). *Symbol and myth: Humbert de Superville's essay on absolute signs in art.* London: Associated Universities Press.

Stevenson, M. B., & Friedman, S. L. (1986). Developmental changes in the understanding of pictorial representations of sound. *Developmental Psychology, 22,* 686–690.

Valdez, P., & Mehrabian, A. (1994). Effects of colors on emotions. *Journal of Experimental Psychology: General, 123,* 394–409.

Verbrugge, R. R., & McCarrell, N. S. (1977). Metaphoric comprehension: Studies in reminding and resembling. *Cognitive Psychology, 9,* 494–533.

Wagner, S., Winner, E., Cicchetti, D., & Gardner, H. (1981). "Metaphorical" mapping in human infants. *Child Development, 52,* 728–731.

Walker, P., & Smith, S. (1984). Stroop interference based on the synaesthetic qualities of auditory pitch. *Perception, 13,* 75–81.

Wapner, S., Werner, H., & Chandler, K. A. (1951). Experiments on sensory-tonic field theory of perception: I. Effect of extraneous stimulation on the visual perception of verticality. *Journal of Experimental Psychology, 42,* 341–345.

Werner, H. (1934). L'unité des sens [The unity of the senses]. *Journal de Psychologie Normale et Pathologique, 31,* 190–205.

Werner, H. (1957). *Comparative psychology of mental development* (Rev. ed.). New York: International Universities Press.

Werner, H. (1963). On expressive language. In G. Kepes (Ed.), *The new landscape in art and science* (pp. 280–282). Chicago: Theobald.

Werner, H., & Wapner, S. (1949). Sensory-tonic field theory of perception. *Journal of Personality, 18,* 88–107.

Werner, H., & Wapner, S. (1952). Toward a general theory of perception. *Psychological Review, 59,* 324–338.

Wright, B., & Rainwater, L. (1962). The meanings of colors. *Journal of General Psychology, 67,* 89–99.

METAPHOR AND SYMBOLIC ACTIVITY *11*(1), 67–84
Copyright © 1996, Lawrence Erlbaum Associates, Inc.

Neurocognitive Mechanisms Underlying Metaphor Comprehension and Other Figurative Language

Curt Burgess
University of California, Riverside

Christine Chiarello
Syracuse University

A complete neurolinguistic model of language comprehension must include the representational and processing components housed in both the left and right cerebral hemispheres. This article focuses on figurative language, most notably the comprehension of metaphors and idioms. We propose that a theory of figurative language relies on a dynamic interplay of bottom-up semantic activation of word information and the top-down constraints offered by the comprehenders' understanding of the conversational pragmatics. We conclude that related language phenomena (indirect speech acts, inferences, proverbs, humor, and some prosodic cues) rely on similar mechanisms that are also shared with literal language comprehension.

The last 20 years has seen the attainment of a notable degree of neurolinguistic sophistication beyond the stereotype that (right-handed) humans are strongly lateralized for language in the left hemisphere (LH) and for visuospatial processing in the right hemisphere (RH). Although there is a consensus that language production is strongly left lateralized, it is now less controversial to acknowledge that the RH possesses certain language comprehension abilities. An increasing amount of evidence suggests that both the RH and LH contribute to the comprehension of semantic rela-

Requests for reprints should be sent to Curt Burgess, Department of Psychology, 1419 Life Sciences Building, University of California, Riverside, Riverside, CA 92521–0426. E-mail: curt@cassandra.ucr.edu.

tions[1] (Beeman et al., 1994; Brownell, 1988; Burgess & Simpson, 1988a, 1988b; Chiarello, 1985, 1988, 1991; Van Lancker, 1987; Zaidel, 1978). For example, both hemispheres show evidence of semantic priming with words that are associatively and semantically related (e.g., *doctor–nurse;* Chiarello, Burgess, Richards, & Pollock, 1990). In fact, the RH actually shows an advantage over the LH in priming for words that have weak or indirect semantic relations (e.g., *dull–moody, lawyer–nurse;* Burgess & Rosen, 1988; Chiarello et al., 1990). Similarly, although dominant and subordinate meanings of ambiguous words show priming in either hemisphere, the subordinate meanings stay active longer in the RH (Burgess, Atchley, Audet, & Arambel, 1994; Burgess & Simpson, 1988a; Chiarello, Maxfield, & Kahan, in press). Thus, both hemispheres contribute to the lexical–semantic processing required by the language comprehension system.

A more specialized account of hemispheric capability emerges, however, when the language subsystems that underlie the processing of syntactic structure are considered in comparison with figurative processing. The domain of syntactic processing appears to be more clearly an LH function. For example, patients with left-hemisphere damage (LHD) seem unable to distinguish between such sentences as *The boy kisses the girl* and *The girl kisses the boy*—a distinction that requires the coordination of the meaning of individual lexical units with the serial syntactic structure of the sentence (Gazzaniga & Hillyard, 1971; Heeschen & Jurgens, 1977). Syntactic comprehension difficulties with LHD patients are most pronounced when the context does not provide any semantic or discourse level constraints for syntactic interpretation (Caplan & Hildebrandt, 1988; Caramazza & Berndt, 1978; Caramazza & Zurif, 1976).

Conversely, more indirect forms of language including figurative language seem most impaired with patients who experience right-hemisphere damage (RHD). For example, these patients have difficulty appreciating the connotative relations between words (Brownell, Potter, Michelow, & Gardner, 1984), making inferences (Beeman, 1993), understanding indirect speech acts (Foldi, 1987; Hirst, LeDoux, & Stein, 1984), appreciating humor (Bihrle, Brownell, Powelson, & Gardner, 1986; Wapner, Hamby, & Gardner, 1981),

---

[1]We focus here on the two major methodologies that have been used to infer complementary hemispheric specializations. One method, used with neurologically normal participants, involves the presentation of stimuli to one visual half field. The anatomy of the visual pathways insures that items briefly presented to one half field are directly transmitted to the opposite hemisphere (e.g., from left visual field to RH). Differential processing of stimuli presented to left and right visual fields allows us to infer hemisphere differences for the process(es) being measured. The second method requires testing of patients with unilateral injury to the LH or RH. Patients with unilateral RH injury, for example, will show deficits on tasks for which the RH is specialized, and their behavior may reflect a greater than normal contribution from the intact LH.

comprehending metaphors (Brownell, 1988), and understanding idiomatic phrases (Kempler, Van Lancker, & Hadler, 1984; Myers & Linebaugh, 1981).

It is now established that a complete neurolinguistic model of language comprehension must include the representational and processing components housed in the left and right cerebral hemispheres (Burgess & Simpson, 1988b; Chiarello, 1991; Millar & Whitaker, 1983; Zaidel, 1978, 1983). Likewise, a complete model of language comprehension must explain how both literal and figurative language can be understood. It is from this distinction between literal and figurative language that many of the more recent contributions about LH and RH language processing have emerged.

Making a distinction between literal and figurative language is fraught with problems: How does the meaning of a sentence differ from the meaning of an utterance (Grice, 1975; Searle, 1975)? Does literal understanding require different underlying semantics than figurative understanding (Burgess & Lund, in press; Glucksberg & Keysar, 1990; Katz & Fodor, 1963; Ortony, 1993)? Must some attempt at literal understanding precede figurative understanding (Gibbs, 1983)? Addressing these questions has involved two basic approaches to the problem of how figurative language is understood. Some investigators have focused on the semantic representations: how they are processed and how this (more or less) bottom-up memory-based approach can resolve the literal–figurative distinction (Glucksberg & Keysar, 1990; Kemper, 1989; Tourangeau & Sternberg, 1982). Others have evaluated the processing of figurative speech in the context of top-down conversational pragmatics. For example, Searle (1993) explicated a set of conversational principles that the comprehension of a metaphor exploits. Pragmatic principles (e.g., cooperativeness, felicity) that facilitate figurative language comprehension could, presumably, be the same as those that underlie general discourse comprehension (Clark & Marshall, 1981). Our categorizing these investigators' theories into bottom-up versus top-down classes of theories does not do justice to the positions they have articulated, nevertheless, theories of figurative language tend to have either a pragmatic (top-down) or semantic (bottom-up) bias.

In this article, we propose that a complete theory of figurative language relies on a dynamic interplay of bottom-up semantic activation of word features and top-down constraints offered by the comprehenders' understanding of the pragmatics inherent in their conversational situation. Furthermore, our goal is to review the literature to demonstrate that each hemisphere makes a unique contribution in this comprehension process. We focus on metaphors and idioms, although we discuss briefly how other related language phenomena (indirect speech acts, proverbs, humor, and prosodic cues) rely on similar mechanisms. We conclude that these cerebral asymmetries in bottom-up activation and top-down constraints are mechanisms that are shared with literal language comprehension. Although there are specializations of function, analysis suggests that Rumelhart (1979) was correct in concluding that the interpretation process for literal and figurative language is the same even though the outcome is different.

CEREBRAL ASYMMETRIES AND METAPHOR COMPREHENSION

One of the earlier observations implicating the RH in metaphor comprehension is due to Weinstein (1964), who noticed that perceptual and emotional processes that are often part of metaphoric communication seemed impaired in patients with RH brain damage. A more systematic approach was taken by McIntyre, Pritchard, and Lombroso (1976). One test they administered to left and right temporal lobe epileptics was the Davitz–Mattis Metaphor Test. Left temporal lobe patients were more inclined to make metaphoric interpretations of the stimuli, which were emotionally evocative scenes. McIntyre et al. argued that the RH compensates with its metaphoric ability for the impaired LH.

A similar set of results was obtained by Winner and Gardner (1977), who had patients with LHD and with RHD select the most appropriate picture to match a verbal metaphor. Winner and Gardner compared psychological–physical metaphors *("A heavy heart can really make a difference")* and cross-sensory metaphors *("It was such colorful music").* The target pictures depicted a metaphoric interpretation of the statement (a person crying in the case of the heavy heart example), a literal depiction (a person struggling with a large, heavy heart), or two foils related to either the noun or adjectival meaning (a red heart or large weight, respectively). The results for both types of metaphors were similar. Compared with the patients with RHD, the patients with LHD more frequently made appropriate metaphoric interpretations and were less likely to select the literally related foils. Patients with RHD were as likely to choose literal interpretations as they were to choose metaphoric interpretations, and unlike the normal patients or patients with LHD they did not find the literal interpretations humorous.

These previous experiments involved metaphors whose interpretations involved computing a meaning from the combination of at least two words (e.g., *heavy heart*). The RH preference for metaphor comprehension generalizes to metaphoric meanings of single words as well. Brownell and his colleagues (Brownell et al., 1984; Brownell, Simpson, Bihrle, Potter, & Gardner, 1990; for a review, see Brownell, 1988) investigated words whose meanings can be metaphoric as well as literal. For example, *cold* and *warm* can metaphorically refer to personality traits through their connotation, or they can denotatively refer to temperature. In Brownell et al. (1984), patients with LHD and RHD were presented three words and asked to pick the two words that were most similar in meaning. The two conditions of most interest in this experiment are the metaphoric-related and the antonymic-related items. The patients with RHD less often selected metaphorically related pairs such as *warm–loving* (achieved by pairing these items less frequently), than did either the patients with LHD or the normal control participants. Conversely, items that were related through their denotation (antonyms) such as *warm–cold,* were judged less similar by the patients with LHD than the patients with RHD or the normal controls. Thus, in this experiment, there is an RH preference for processing single-word metaphoric meanings, which extends the basic result found by Winner and Gardner (1977) with adjective–noun metaphors.

Although the pattern of results is similar in indicating an RH preference for making metaphoric interpretations, the Brownell et al. (1984) single-word stimuli were also lexical ambiguities. It may be that the combinatorial processes involved in interpreting multiword metaphors are different from those used for single-word metaphors, which are lexically ambiguous. Burgess and Simpson (1988b) raised this question in noting that the metaphorical meanings of the examples in Brownell et al. (1984) corresponded to their less frequent (subordinate) meanings. Burgess and Simpson (1988a; see also Burgess et al., 1994) showed that the retrieval of subordinate word meanings is protracted over time in the RH compared to the LH. Brownell et al. (1990) directly addressed this concern in a follow-up to their original experiment. As before, they had patients select the two words that were most similar. An experimental trial included the metaphor target, for example, *deep,* and a metaphoric synonym, *wise,* or a literal foil, *lake.* Nonmetaphoric ambiguous words included items such as the target, *suit,* with the subordinate synonym, *trial,* and the foil, *tailor.* The pattern of results in this experiment paralleled their previous experiment; patients with RHD made fewer metaphorically related similarity matches than the patients with LHD. These two patient groups did not differ on the nonmetaphoric ambiguous nouns, however, and performance in this condition was better than on the metaphor trials. There are two possible explanations for the lack of asymmetry in the ambiguity condition. One is that the offline nature of the task may have obscured asymmetries that are observed at earlier stages of processing. Another possibility is that the ambiguous targets were synonyms of the ambiguous word (e.g., *pupil–student*). With virtually complete semantic overlap, facilitation with these items should be bilateral (see Chiarello et al., 1990). In either case, however, there is an important role for the RH in the interpretation of metaphor, whether as single words or phrasal units.

Another experiment that investigated neuropsychological aspects of metaphor comprehension is worth noting due to its theoretical convergence and unique methodology. Bottini et al. (1994) investigated metaphor and literal sentence processing with a group of normal, intact participants by looking at relative regional cerebral blood flow using positron emission tomography (PET). The PET procedure reveals regional differences in brain activation during the performance of cognitive tasks. Participants were presented three kinds of stimuli during the experiment: metaphors (either plausible or implausible); literal sentences (either plausible or implausible); and sentence-like strings of words, half of which contained nonwords. Several examples follow:

1a. *"The investors were squirrels collecting nuts."* [plausible metaphor]
1b. *"The investors were trams."* [implausible metaphor]

2a. *The boy used stones as paperweights.* [plausible sentence]
2b. *Tim used feathers as paperweights.* [implausible sentence]

The crucial result for our discussion is that, compared with literal sentences, metaphoric sentences generated increased activity in six regions in the RH. An increase in LH processing was detected for the literal sentences compared with the single-word strings. Although this experiment does not rule out the possibility that more complex metaphors might recruit LH resources during the comprehension process, the pattern of the PET results show the same RH advantage for the metaphors as Brownell and others (Brownell, 1988; Brownell et al., 1984; Brownell et al., 1990; Winner & Gardner, 1977). This experiment is important beyond its ability to more finely pinpoint neuroanatomical processing locations; the stimuli were more complex, being full sentences instead of single words with metaphoric meanings or simple adjective–noun combinations.

CEREBRAL ASYMMETRIES AND IDIOM COMPREHENSION

In the preceding section, we presented evidence suggesting that the RH plays an important role in the comprehension of metaphors. In this section, we will consider neurolinguistic studies of idiom comprehension in some detail. Our discussion will focus specifically on the nature of hemispheric asymmetries in the processing of literal and idiomatic language.

An experiment by Van Lancker and Kempler (1987) explored the neural underpinnings of figurative and literal comprehension using a methodology similar to that of Winner and Gardner (1977). They had patients with LHD and patients with RHD select the appropriate line drawing (out of a set of four) that corresponded to either a single concrete noun *(bird),* a sentence containing a figurative phrase *("He's turning over a new leaf"),* or a novel literal sentence *(He's sitting deep in the bubbles)*. For the idioms, one drawing was related to the literal interpretation of the idiom and another, to the figurative interpretation of the idiom. The other two drawings depicted a scene that was either opposite in meaning to the idiom or was irrelevant. The drawings used with the novel literal sentences were somewhat different. All three distractors for these sentences involved some change in grammatical or thematic roles of the nouns (e.g., agent–patient reversal) or misassignment of adjective modification (e.g., the sentence, *When the happy girl pushes, the angry boy swings,* had a drawing of an unhappy girl pushing a boy). Correct performance on the literal sentences, therefore, required careful attention to both the semantic and syntactic details of the sentences. A critical finding from the experiment was an interaction between patient type and stimulus set. Specifically, patients with LHD performed worse than patients with RHD on the novel literal sentences (those that required detailed syntactic analysis) but outperformed the patients with RHD on the figurative sentences. Apparently, LHD interferes significantly with syntactic analysis but is less disruptive of figurative processing. Based on these results, Van Lancker and Kempler argued that syntactic analysis is not a necessary component of idiomatic interpretation. They suggested that the results support an idiom-as-a-word hypothesis or what Swinney and Cutler (1979) refer to as the lexical representation hypothesis. According to this view, an idiom's meaning is directly (and holistically) retrieved from memory (presumably from the RH) without any consideration of the idiom's internal syntactic structure.

The assumption, however, that syntactic processing is not actively engaged during the reading of an idiom is not supported by the results of Peterson and Burgess (1993). They demonstrated that the magnitude of syntactic priming that occurs with idiomatic strings is equal to (or even surpasses) that found with literal phrases. This result suggests that syntactic processing occurs in an idiomatic phrase as it would in a literal construction. Perhaps analysis of the figurative meaning of an idiom occurs in the RH and is neurally dissociated from LH parsing.

If this is accurate, then an alternative to the idiom-as-a-word hypothesis can be put forth for Van Lancker and Kempler's (1987) finding that patients with LHD (who presumably have impaired syntactic capabilities) performed better on idiomatic sentences than patients RHD (who have intact parsers). Two factors may have conspired to produce this result. First, it may be that the patients with RHD had difficulty with the idiomatic strings due to a particular processing bias of the LH. Specifically, the LH has been characterized as making rapid and inflexible commitments when interpreting lexical items (Brownell, Potter, Bihrle, & Gardner, 1986; Burgess & Simpson, 1988b; Chiarello, 1991). If so, the rapid commitment to an interpretation is more likely to be associated with a literal interpretation; that is, patients with RHD may have interpreted the initial words of the idiom literally and, as a result, may have had trouble appreciating the figurative status of the remainder of the phrase. According to this view, the LH has difficulty with idiomatic phrases because it is "garden-pathed" away from a figurative interpretation during processing.

This explanation is qualitatively different from the account given by the idiom-as-a-word hypothesis. According to a neural perspective on that hypothesis, patients with RHD have difficulty with idiomatic phrases because idiomatic entries in the lexicon occur only in the RH. The obvious problem with this hypothesis is that there is no a priori account of why the LH is unable to represent "idiomatic" words. Our account, in contrast, focuses not on representational differences between the hemispheres but on well-documented processing differences. As a result, our hypothesis fits more directly into extant neurolinguistic models of comprehension.

Although this hypothesis accounts for the poor performance by patients with RHD on the idiomatic phrases, it does not explain why the patients with LHD (whose parsers were apparently impaired) were able to comprehend the idioms at a relatively high level (although at a level still below that of normal subjects). Research by Caramazza and Zurif (1976) may provide an answer to this question, which suggests a second factor that contributes to why the patients with LHD in Van Lancker and Kempler's (1987) study performed better on the idiomatic sentences than patients with RHD. These investigators demonstrated that, although patients with LHD have difficulty interpreting sentences like *The tiger is chasing a fat lion,* they performed much better on sentences like *The bicycle that the boy is holding is broken.* The critical difference between these sentences appears to be that, in the first, the subject and object noun phrases are reversible (in the sense that either could pragmatically be the agent of the verb), whereas in the second sentence, the nouns are nonreversible (because *bicycle* is an inanimate object and thus cannot serve as the agent of the sentence). These data suggest that the RH is adept at utilizing nonsyntactic cues to derive the structural relations among the components of a sentence. Perhaps such analyses were used by patients with LHD in the Van Lancker and Kempler study in order to determine the structure of idiomatic phrases as well. Consider, for example, how the RH might analyze a sentence like *He kicked the bucket.* Because the subject and object noun phrases of this sentence are literally nonreversible (i.e., *bucket* cannot be the agent of *kicked* because it is an inanimate noun), the RH could presumably determine the underlying (literal) structure of this sentence without syntactic mediation because the frozen structure would likely be mapped onto the idiomatic meaning. The retrieval of this structure might subsequently allow access to the idiom's representation in memory (which itself contains the information that *bucket* is the underlying object of the phrase). According to this view, then, the RH might sometimes be able to "bootstrap" its way to a figurative interpretation by first deriving a thematic representation of an idiomatic phrase. This bootstrapping hypothesis predicts that patients with LHD might be particularly sensitive to the literal relations among the components of an idiomatic phrase.

In general, the basic finding by Van Lancker and Kempler (1987) that patients with RHD were more impaired on the idiomatic stimuli than were the patients with LHD is consistent with the metaphor results discussed earlier. Furthermore, Myers and Linebaugh (1981) conducted a very similar study wherein patients selected pictures that corresponded to literal, idiomatic, or control interpretations of a context paragraph and found similar results. Their experiment is notable in that a context-rich paragraph was provided that supported an idiomatic interpretation, thus ruling out the possibility that an impoverished context resulted in the deficit shown by the patients with RHD.

Perhaps it could now be concluded that the results from the idiom literature mirror those described previously from the metaphor research. However, a recent article by Tompkins, Boada, and McGarry (1992) suggests that any such conclusion would be premature. These researchers made use of a distinction in cognitive psychology (Neely, 1977; Posner & Snyder, 1975) and in cognitive neuropsychology (Blumstein, Milberg, & Shrier, 1982; Burgess & Simpson, 1988a; Chiarello, 1985; Milberg & Blumstein, 1981) between automatic and controlled processing. Processing is considered to be automatic if attention is not required and, at least with word recognition, typically corresponds to a perceptually driven bottom-up spread of activation in memory. Controlled processing involves an attentional component and more central cognitive processes and thus is considered top-down. Idiom experiments in the neuropsychological literature tended to utilize tasks such as text–picture matching (Myers & Linebaugh, 1981; Van Lancker & Kempler, 1987). Such a task would be considered offline; that is, the ultimate matching decision made is the outcome of a complex sequence of cognitive processes and necessarily has a strong attentional, top-down component. Most psycholinguistic experiments with idioms have used procedures that would be online, such as cross-modal priming in which participants engage in a task that is incidental to the overall comprehension process under study (e.g., Cacciari & Tabossi, 1988). For example, an online task like making a lexical decision to a word target related to an idiomatic meaning is less likely to involve centrally driven strategies than an offline task that requires an overt decision about the relatedness of an idiom and a picture.

This is the background for the Tompkins et al. (1992) experiment. They argued against the idea that the RH advantage for idiom comprehension reflects early stages of language comprehension in which the retrieval of the idiomatic meaning may be taking place.[2] To address this concern, Tompkins et al. conducted an online word-monitoring task and an offline idiom-definition task. The online task consisted of the detection of a target word in an idiom embedded in the last sentence of a two-sentence "story." The target word on a critical trial was the last word of the idiom (e.g., "smelled a *rat*"). The target words also occurred in nonidiomatic control sentences. The offline task was straightforward: Patients were simply asked to explain in a few words what each idiom meant. This particular task was selected because the authors thought that the text–picture matching procedure might be too difficult and would therefore result in an underestimation of the patients' sensitivity to the idiomatic choices.

Both patients with LHD and patients with RHD performed similarly to normal controls on the online task; that is, responses to target words in idiomatic contexts were faster than when targets were in the nonidiomatic control condition. Tompkins et al.'s (1992) interpretation of this result was that the initial retrieval of the idiomatic meaning is not impaired with either the patients with LHD or the patients with RHD. The offline results showed a different pattern of results, however. Patients with LHD and patients with RHD were worse than normal controls at defining the idiom and did not differ from each other in this regard. Presumably this shows that, in understanding an idiom, the difficulty for both groups of patients occurs at a later stage in language comprehension than the results from Myers and Linebaugh (1981) and Van Lancker and Kempler (1987) research would suggest. Tompkin et al.'s rationale for why the offline task did not replicate the RH advantage for idiom comprehension was that this asymmetry emerges with the more difficult processing associated with the text–picture matching task. This hypothesis needs further substantiation, but differences between Tompkins et al.'s results and the other idiom experiments cited suggest that the two cerebral hemispheres make different contributions to idiom comprehension.

THEORIES OF FIGURATIVE LANGUAGE: BOTTOM UP AND TOP DOWN

In this section, we consider how any complete theory of figurative language will require both bottom-up and top-down components. We know from word recognition and priming research that cerebral asymmetries are most likely to be evident in top-down conditions that encourage controlled, atten-

[2]This claim needs to be taken seriously, given that the bulk of the psycholinguistic literature suggests that idiom comprehension occurs very rapidly, automatically, and in parallel with or even prior to a literal computation (Estill & Kemper, 1982; Gibbs, 1986; Ortony, Shallert, Reynolds, & Antos, 1978; Peterson & Burgess, 1993; Swinney & Cutler, 1979).

tional processing (e.g., tested with the use of long prime durations or a high proportion of "related" prime-target trials). Likewise, at earlier stages of word recognition (e.g., tested with the use of brief prime durations or a low proportion of related trials), a common, bottom-up, pattern of results is that asymmetries are not present or are attenuated (Burgess & Rosen, 1988; Burgess & Simpson, 1988a; Chiarello et al., 1990). The two hemispheres are similar in the semantic representations that are activated via bottom-up input, although the RH has a broader range of semantic activation (Beeman et al., 1994; Burgess & Simpson, 1988a; Chiarello et al., 1990). For example, Chiarello, Senehi, and Nuding (1987) found the differential priming of abstract and concrete words to occur only in a controlled priming condition in which there was a high probability of related trials. When the abstract and concrete related pairs constituted a low proportion of related items, the magnitude of priming was equal across the visual fields. Along similar lines, Burgess and colleagues (Burgess et al., 1994; Burgess & Simpson, 1988a) found that the greatest processing asymmetries during the retrieval of ambiguous word meanings occur later in processing when greater attentional resources can be allocated.

It seems clear that mapping out the time course of memory activation and distinguishing between automatic and controlled processes have contributed an important aspect to our understanding of the nature of cerebral asymmetries of word recognition. This implies that the time course of meaning activation could be important in retrieving the meanings of single-word metaphors. One prediction would be that single-word metaphoric meanings like those investigated by Brownell and colleagues (Brownell et al., 1984, Brownell et al., 1990), would be available in the LH at an early stage of processing prior to the operation of attentional, top-down processing. With sufficient supporting context for the metaphoric interpretation of a word, it seems plausible that metaphoric comprehension could be subserved by the LH. It seems just as likely that time course and, by extension, the distinction between bottom-up and top-down processing would be important in the processing of idiomatic language as well. Cacciari and Tabossi (1988) showed that the point during the presentation of an idiom at which meaning retrieval occurs is a function of the predictability of the idiom. Readers are more sensitive to the statistical likelihood of particular upcoming words in an idiom than in a nonidiomatic phrase. Such expectations illustrate top-down constraints on processing. In general, it would seem that to fully understand cerebral asymmetries in language processing, the role of both top-down and bottom-up components should be taken into account.

However, our earlier assertion was that a complete theory of figurative language relies on a dynamic interplay of bottom-up semantic activation of word information and the top-down constraints offered by the comprehender's understanding of the conversational situation. From the very beginning of psycholinguistic investigations of figurative language, the speaker's intention and the comprehender's understanding of that intention have played an important role in theory development. In fact, for Grice (1975) and Searle (1975) it is crucial. Grice maintains that efficient communication requires adherence to the cooperative principle. Part of this involves that the utterance be true and relevant to the goal of the conversation. Intention of the speaker is important as well. Searle (1969; see also Austin, 1962) articulated a set of social conventions that speakers and listeners presume to be operative during a conversation. These can include the speaker understanding a listener's ability to do something and the listener understanding the speaker's desire for an action or for something to be understood. More recently, Clark and Marshall (1981) discussed similar issues about how speakers and listeners determine intention: They refer to this as *mutual knowledge*. Inadequacies in mutual knowledge can foster a lapse in communication. More generally, such information is known as pragmatics or a person's knowledge about the world, both with respect to the social–interpersonal situation as well as more general knowledge.

Winner, Brownell, Blum, and Pincus (1994) recently evaluated the effect of ability of patients with RHD to make correct inferences about a speaker's intention when uttering an untrue statement. There were two types of untrue statements, lies and jokes, read by patients and normal controls. When a story was read and the critical utterance was understood to be a lie, the comprehender knew that the speaker's intent was to deceive. However, when the story was read and the critical utterance was understood to be an ironic joke, there was no intention to deceive. Both patients with RHD and normal controls were able to understand that a liar expects to be believed but that a joker with humorous intent does not. The crucial difference between the patients with RHD and the control participants was that the patients with RHD had difficulty making the inferences required to distinguish between the speaker intending the utterance as a joke or a lie.

A somewhat analogous result is that patients with RHD have difficulty interpreting indirect speech acts. They tend to interpret them literally, unlike patients with LHD who have much less difficulty understanding the conveyed meaning (Foldi, 1987; Hirst et al., 1984; Stemmer, Giroux, & Joanette, 1994). It would seem that the patients with RHD, in part, are deficient in the higher level knowledge needed to understand the social conventions required to correctly make this interpretation. Of course, one cannot be completely sure, given the findings of Tompkins et al. (1992), that metaphor comprehension was equally impaired in patients with RHD and patients with LHD when tested in an automatic-processing condition. It remains to be seen if there are early stages of indirect speech act comprehension such as initial, multiple interpretations that are available but not utilized. Stage considerations aside, it remains clear that the RH has a role in the comprehension of indirect speech acts.

Both the participation of the RH in understanding intent and the deficit in the necessary inference making demonstrated by the patients with RHD suggest that pragmatics may be an important constraint operating on figurative language processing in general (Stemmer et al., 1994; Winner et al., 1994). Such a conclusion is similar to the original proposals of Searle (1969) and Grice (1975), who discussed the importance of cooperation and intention in avoiding misunderstanding during conversation.

CONCLUSIONS

One general finding from the research on idioms and metaphors is that both figurative speech acts are understood more quickly and efficiently when there are top-down effects of supporting context (Cacciari & Tabossi, 1988). Otherwise, the comprehension system relies more on bottom-up, word-level input to resolve the statement. The same relation between supporting context and ease of comprehension can be found with nonfigurative speech, such as lexical ambiguity (Burgess, Tanenhaus, & Seidenberg, 1989; Simpson & Krueger, 1991; Tabossi, 1988; Tabossi & Zardon, 1993) and syntactic ambiguity (Burgess, Tanenhaus, & Hoffman, 1994; MacDonald, Pearlmutter, & Seidenberg, 1994; Trueswell, Tanenhaus, & Garnsey, 1994). We reviewed evidence showing that the RH makes an important contribution to inferencing—an important component in the development of a mental model of the relation between the social environment and figurative utterances. We also discussed literature that illustrates some differences in how the two hemispheres initially retrieve information. Although this initial process is rapid and focused in the LH, the RH activates information more slowly and does not actively select information, producing a broader field of semantic activation than the LH (Burgess & Simpson, 1988a, 1988b; Chiarello, 1985; Chiarello & Richards, 1992). Such RH processes, namely, activating a broader range of semantic information and increasing sensitivity to pragmatic information, likely play an important role in understanding figurative language. Thus, the two hemispheres comprise a system whereby both hemispheres demonstrate a bottom-up activation of word meanings, but the RH shows a broader range of activation. We suspect that the broader RH semantic activation is crucial in understanding metaphors, particularly when the metaphor is not supported well by context.

Both hemispheres likewise use top-down processes. In the LH, these top-down processes enable rapid semantic and syntactic commitments during comprehension. In the RH, top-down processes are important for pragmatic interpretation such as inferencing and interpreting the speaker's mood and intention. Because both pragmatic and broad-based semantic processes are important for comprehending figurative language, there is a special role for the RH in language comprehension. Yet, LH processes may also be needed for aspects of figurative language comprehension, especially those involved in syntactic analysis of idioms (Peterson & Burgess, 1993) or when ample context precedes the figurative expression. We suggest that future neuropsychological models of figurative language should acknowledge that top-down constraints of context (general knowledge and pragmatic information) can interact with bottom-up information that is rapidly and automatically made available during initial word recognition. Furthermore, understanding this relation between bottom-up and top-down processing requires methodologies that increasingly separate automatic- and controlled-processing components.

Although this article has focused on issues in metaphor and idiom comprehension, similar issues are present for other types of figurative speech acts. There is evidence that the RH is involved in inference making that contributes to discourse coherence (Beeman, 1993; Brownell, Carroll, Rehak, & Wingfield, 1992; Brownell et al., 1986; cf. McDonald & Wales, 1986; Tompkins & Mateer, 1985), comprehension of humor (Bihrle et al., 1986; Gardner, Ling, Flamm, & Silverman, 1975), and indirect speech acts (Foldi, 1987; Foldi, Cicone, & Gardner, 1983; Hirst et al., 1984; Stemmer et al., 1994; Wilcox, Davis, & Leonard, 1978). In addition, proverb interpretation (Van Lancker, 1990; see Honeck & Temple, 1994, for a review of proverbs in the context of automatic and controlled processing) and the use of emotional prosodic cues (Kreiman & Van Lancker, 1988; Van Lancker, Cummings, Kreiman, & Dobkin, 1988; Van Lancker, Kreiman, & Cummings, 1989; cf. Emmorey, 1987; Ouellette & Baum, 1993) also seem to engage the RH.

The broad view suggested by neurological evidence is that a strict dichotomy of figurative and literal language is likely false, as Rumelhart (1979; see also Lakoff, 1994) concluded in his review of psychological evidence. However, it seems apparent that without the contribution of the RH, language would be a greatly impoverished capability.

ACKNOWLEDGMENTS

This research was supported by NSF Presidential Faculty Fellow Award SBR–9453406 to Curt Burgess. We thank Chad Audet for his bibliographic assistance and Catherine Decker and Richard Honeck for their helpful comments and criticisms.

REFERENCES

Austin, J. L. (1962). *How to do things with words.* Oxford, England: Oxford University Press.

Beeman, M. (1993). Semantic processing in the right hemisphere may contribute to drawing inferences from discourse. *Brain and Language, 44,* 80–120.

Beeman, M., Friedman, R., Grafman, J., Perez, E., Diamond, S., & Lindsay, M. (1994). Summation priming and coarse semantic coding in the right hemisphere. *Journal of Cognitive Neuroscience, 6,* 26–45.

Bihrle, A. M., Brownell, H. H., Powelson, J. A., & Gardner, H. (1986). Comprehension of humorous and nonhumorous materials by left and right brain-damaged patients. *Brain and Cognition, 5,* 399–411.

Blumstein, S. E., Milberg, W., & Shrier, R. (1982). Semantic processing in aphasia: Evidence from an auditory lexical decision task. *Brain and Language, 17,* 301–315.

Bottini, G., Corcoran, R., Sterzi, R., Paulesu, E., Schenone, P., Scarpa, P., Frackowiak, R. S. J., & Frith, C. D. (1994). The role of the right hemisphere in the interpretation of figurative aspects of language: A positron emission tomography activation study. *Brain, 117,* 1241–1253.

Brownell, H. H. (1988). Appreciation of metaphoric and connotative word meaning by brain-damaged patients. In C. Chiarello (Ed.), *Right hemisphere contributions to lexical semantics* (pp. 19–31). Berlin: Springer-Verlag.

Brownell, H. H., Carroll, J. J., Rehak, A., & Wingfield, A. (1992). The use of pronoun anaphora and speaker mood in the interpretation of conversational utterances by right hemisphere brain-damaged patients. *Brain and Language, 43,* 121–147.

Brownell, H. H., Potter, H. H., Bihrle, A. M., & Gardner, H. (1986). Inference deficits in right brain-damaged patients. *Brain and Language, 27,* 310–321.

Brownell, H. H., Potter, H. H., Michelow, D., & Gardner, H. (1984). Sensitivity to lexical denotation and connotation in brain-damaged patients: A double dissociation? *Brain and Language, 22,* 253–265.

Brownell, H. H., Simpson, T. L., Bihrle, A. M., Potter, H. H., & Gardner, H. (1990). Appreciation of metaphoric alternative word meanings by left and right brain-damaged patients. *Neuropsychologia, 28,* 375–383.

Burgess, C., Atchley, R. A., Audet, C., & Arambel, S. (1994, November). *Cerebral hemispheric mechanisms of lexical ambiguity: Dominance and sentence context effects.* Paper presented at the meeting of the Psychonomics Society, St. Louis, MO.

Burgess, C., & Lund, K. (in press). Modeling the cerebral asymmetries of semantic memory in high-dimensional space: HAL's brain [Abstract]. *Brain and Cognition.*

Burgess, C., & Rosen, A. (1988, April). *Priming of emotional words in the cerebral hemispheres.* Paper presented at the Linguistic Studies Conference on Language and Communication, Syracuse University, Syracuse, NY.

Burgess, C., & Simpson, G. B. (1988a). Cerebral hemispheric mechanisms in the retrieval of ambiguous word meanings. *Brain and Language, 33,* 86–104.

Burgess, C., & Simpson, G. B. (1988b). Neuropsychology of lexical ambiguity resolution: The contribution of divided visual field studies. In S. L. Small, G. W. Cottrell, & M. K. Tanenhaus (Eds.), *Lexical ambiguity resolution: Perspectives from psycholinguistics, neuropsychology, and artificial intelligence* (pp. 411–430). San Mateo, CA: Morgan Kaufmann.

Burgess, C., Tanenhaus, M. K., & Hoffman, M. (1994). Parafoveal and semantic effects on syntactic ambiguity resolution. *Proceedings of the Cognitive Science Society* (pp. 96–99). Hillsdale, NJ: Lawrence Erlbaum Associates, Inc.

Burgess, C., Tanenhaus, M. K., & Seidenberg, M. (1989). Context and lexical access: Implications of nonword interference for lexical ambiguity resolution. *Journal of Experimental Psychology: Learning, Memory and Cognition, 15,* 620–632.

Cacciari, C., & Tabossi, P. (1988). The comprehension of idioms. *Journal of Memory and Language, 27,* 668–683.

Caplan, D., & Hildebrandt, N. (1988). *Disorders of syntactic comprehension.* Cambridge, MA: MIT Press.

Caramazza, A., & Berndt, R. S. (1978). Semantic and syntactic processes in aphasia: A review of the literature. *Psychological Bulletin, 85,* 898–918.

Caramazza, A., & Zurif, E. B. (1976). Dissociation of algorithmic and heuristic processes in language comprehension: Evidence from aphasia. *Brain and Language, 3,* 572–582.

Chiarello, C. (1985). Hemisphere dynamics in lexical access: Automatic and controlled priming. *Brain and Language, 26,* 146–172.

Chiarello, C. (1988). Semantic priming in the intact brain: Separate roles for the right and left hemispheres? In C. Chiarello (Ed.), *Right hemisphere contributions to lexical semantics* (pp. 59–69). Berlin: Springer-Verlag.

Chiarello, C. (1991). Interpretation of word meanings by the cerebral hemispheres: One is not enough. In P. J. Schwanenflugel (Ed.), *The psychology of word meanings* (pp. 251–278). Hillsdale, NJ: Lawrence Erlbaum Associates, Inc.

Chiarello, C., Burgess, C., Richards, L., & Pollock, A. (1990). Semantic and associative priming in the cerebral hemispheres: Some words do, some words don't . . . sometimes, some places. *Brain and Language, 38,* 75–104.

Chiarello, C., Maxfield, L., & Kahan, T. (in press). Initial right hemisphere activation of subordinate word meanings is not due to homotopic callosal inhibition. *Psychonomic Bulletin & Review.*

Chiarello, C., & Richards, L. (1992). Another look at categorical priming in the cerebral hemispheres. *Neuropsychologia, 30,* 381–392.

Chiarello, C., Senehi, J., & Nuding, S. (1987). Semantic priming with abstract and concrete words: Differential asymmetry may be postlexical. *Brain and Language, 31,* 43–60.

Clark, H. H., & Marshall, C. E. (1981). Definite reference and mutual knowledge. In A. K. Joshi, I. Sag, & B. Webber (Eds.), *Elements of discourse understanding* (pp. 10–63). Cambridge, England: Cambridge University Press.

Emmorey, K. D. (1987). The neurological substrates for prosodic aspects of speech. *Brain and Language, 30,* 305–320.

Estill, R., & Kemper, S. (1982). Interpreting idioms. *Journal of Psycholinguistic Research, 6,* 559–568.

Foldi, N. S. (1987). Appreciation of pragmatic interpretations of indirect commands: Comparison of right and left brain-damaged patients. *Brain and Language, 31,* 88–108.

Foldi, N. S., Cicone, M., & Gardner, H. (1983). Pragmatic aspects of communication in brain-damaged patients. In S. Segalowitz (Ed.), *Language functions and language organization* (pp. 71–84). New York: Academic.

Gardner, H., Ling, P. K., Flamm, L., & Silverman, J. (1975). Comprehension and appreciation of humorous material following brain damage. *Brain, 98,* 399–412.

Gazzaniga, M. S., & Hillyard, S. A. (1971). Language and speech capacity of the right hemisphere. *Neuropsychologia, 9,* 273–280.

Gibbs, R. W. (1983). Do people always process the literal meanings of indirect requests? *Journal of Experimental Psychology: Learning, Memory, and Cognition, 9,* 524–533.

Gibbs, R. (1986). Skating on thin ice: Literal meaning and understanding idioms in conversation. *Discourse Processes, 9,* 17–30.

Glucksberg, S., & Keysar, B. (1990). Understanding metaphorical comparisons: Beyond similarity. *Psychological Review, 97,* 3–18.

Grice, H. P. (1975). Logic and conversation. In P. Cole & J. L. Morgan (Eds.), *Syntax and semantics: Vol. 3. Speech acts* (pp. 41–58). New York: Seminar Press.

Heeschen, C., & Jurgens, R. (1977). Pragmatic-semantic and syntactic factors influencing ear differences in dichotic listening. *Cortex, 13,* 74–84.

Hirst, W., LeDoux, J., & Stein, S. (1984). Constraints on the processing of indirect speech acts: Evidence from aphasiology. *Brain and Language, 23,* 26–33.

Honeck, R. P., & Temple, J. G. (1994). Proverbs: The extended conceptual base and great chain metaphor theories. *Metaphor and Symbolic Activity, 9,* 85–112.

Katz, J. J., & Fodor, J. A. (1963). The structure of a semantic theory. *Language, 39,* 170–210.

Kemper, S. (1989). Priming the comprehension of metaphors. *Metaphor and Symbolic Activity, 4,* 1–17.

Kempler, D., Van Lancker, D., & Hadler, B. (1984). *Familiar phrase recognition in brain-damaged and demented adults.* Paper presented at the meeting of the Academy of Aphasia, Los Angeles, CA.

Kreiman, J., & Van Lancker, D. (1988). Hemispheric specialization for voice recognition: Evidence from dichotic listening. *Brain and Language, 34,* 246–252.

Lakoff, G. (1994). What is a conceptual system? In W. F. Overton & D. S. Palermo (Eds.), *The nature and ontogenesis of meaning* (pp. 41–58). Hillsdale, NJ: Lawrence Erlbaum Associates, Inc.

MacDonald, M. C., Pearlmutter, N. J., & Seidenberg, M. S. (1994). The lexical nature of syntactic ambiguity resolution. *Psychological Review, 101,* 676–703.

McDonald, S., & Wales, R. (1986). An investigation of the ability to process inferences in language following right hemisphere brain damage. *Brain and Language, 29,* 68–80.

McIntyre, M., Pritchard, P. B., & Lombroso, C. T. (1976). Left and right temporal lobe epileptics: A controlled investigation of some psychological differences. *Epilepsia, 17,* 377–386.

Milberg, W., & Blumstein, S. E. (1981). Lexical decision and aphasia: Evidence for semantic processing. *Brain and Language, 14,* 371–385.

Millar, J. M., & Whitaker, H. A. (1983). The right hemisphere's contribution to language: A review of the evidence from brain-damaged subjects. In S. J. Segalowitz (Ed.), *Language functions and brain organization* (pp. 87–113). New York: Academic.

Myers, P. S., & Linebaugh, C. W. (1981). Comprehension of idiomatic expressions by right-hemisphere-damaged adults. In R. H. Brookshire (Ed.), *Clinical aphasiology: Conference proceedings.* Minneapolis, MN: BRK.

Neely, J. H. (1977). Semantic priming and retrieval from lexical memory: Roles of inhibitionless spreading activation and limited-capacity attention. *Journal of Experimental Psychology: General, 106,* 226–254.

Ortony, A. (1993). The role of similarity in similes and metaphors. In A. Ortony (Ed.), *Metaphor and thought* (2nd ed., pp. 342–356). New York: Cambridge University Press.

Ortony, A., Shallert, D., Reynolds, R., & Antos, S. (1978). Interpreting metaphors and idioms: Some effects of context on comprehension. *Journal of Verbal Learning and Verbal Behavior, 17,* 465–477.

Ouellette, G. P., & Baum, S. R. (1993). Acoustic analysis of prosodic cues in left- and right-hemisphere-damaged patients. *Aphasiology, 8,* 257–283.

Peterson, R. R., & Burgess, C. (1993). Syntactic and semantic processing during idiom comprehension: Neurolinguistic and psycholinguistic dissociations. In C. Cacciari & P. Tabossi (Eds.), *Idioms: Processing, structure, and interpretation* (pp. 201–225). Hillsdale, NJ: Lawrence Erlbaum Associates, Inc.

Posner, M. I., & Snyder, C. R. R. (1975). Attention and cognitive control. In R. Solso (Ed.), *Information processing and cognition: The Loyola symposium* (pp. 669–682). Hillsdale, NJ: Lawrence Erlbaum Associates, Inc.

Rumelhart, D. E. (1979). Some problems with the notion of literal meanings. In A. Ortony (Ed.), *Metaphor and thought* (pp. 78–90). New York: Cambridge University Press.

Searle, J. R. (1969). *Speech acts.* New York: Academic.

Searle, J. R. (1975). Indirect speech acts. In P. Cole & J. L. Morgan (Eds.), *Syntax and semantics: Vol. 3. Speech acts* (pp. 59–82). New York: Seminar Press.

Searle, J. R. (1993). Metaphor. In A. Ortony (Ed.), *Metaphor and thought* (2nd ed., pp. 83–111). New York: Cambridge University Press.

Simpson, G. B., & Kreuger, M. A. (1991). Selective access of homograph meanings in sentence context. *Journal of Memory and Language, 30,* 627–643.

Stemmer, B., Giroux, F., & Joanette, Y. (1994). Production and evaluation of requests by right hemisphere brain-damaged individuals. *Brain and Language, 47,* 1–31.

Swinney, D. A., & Cutler, A. (1979). The access and processing of idiomatic expressions. *Journal of Verbal Learning and Verbal Behavior, 18,* 523–534.

Tabossi, P. (1988). Accessing lexical ambiguity in different types of sentential contexts. *Journal of Memory and Language, 27,* 324–340.

Tabossi, P., Zardon, F. (1993). The activation of idiom meaning in spoken language comprehension. In C. Cacciari & P. Tabossi (Eds.), Idioms, processing, structure and interpretation (pp. 145–162). Hillsdale, NJ: Lawrence Erlbaum Associates, Inc.

Tompkins, C. A., Boada, R., & McGarry, K. (1992). The access and processing of familiar idioms by brain-damaged and normally aging adults. *Journal of Speech and Hearing Research, 35,* 626–637.

Tompkins, C. A., & Mateer, C. A. (1985). Right hemisphere appreciation of prosodic and linguistic indications of implicit attitude. *Brain and Language, 24,* 185–203.

Tourangeau, R., & Sternberg, R. J. (1982). Understanding and appreciating metaphors. *Cognition, 11,* 203–244.

Trueswell, J. C., Tanenhaus, M. K., & Garnsey, S. M. (1994). Semantic influences on parsing: Use of thematic role information in syntactic ambiguity resolution. *Journal of Memory and Language, 33,* 285–318.

Van Lancker, D. (1987). Nonpropositional speech: Neurolinguistic studies. In A. W. Ellis (Ed.), *Progress in the psychology of language* (Vol. 3, pp. 49–118). London: Lawrence Erlbaum Associates, Inc.

Van Lancker, D. (1990). The neurology of proverbs. *Behavioural Neurology, 3,* 169–187.

Van Lancker, D. R., Cummings, J. L., Kreiman, J., & Dobkin, B. H. (1988). Phonagnosia: A dissociation between familiar and unfamiliar voices. *Cortex, 24,* 195–209.

Van Lancker, D. R., & Kempler, D. (1987). Comprehension of familiar phrases by left- but not by right-hemisphere damaged patients. *Brain and Language, 32,* 265–277.

Van Lancker, D. R., Kreiman, J., & Cummings, J. (1989). Voice perception deficits: Neuroanatomical correlates of phonagnosia. *Journal of Clinical and Experimental Neuropsychology, 11,* 665–674.

Wapner, W., Hamby, S., & Gardner, H. (1981). The role of the right hemisphere in the apprehension of complex linguistic materials. *Brain and Language, 14,* 15–33.

Weinstein, E. A. (1964). Affections of speech with lesions of the non-dominant hemisphere. *Research Publications: Association for Research in Nervous and Mental Disease, 42,* 220–225.

Wilcox, M. J., Davis, G. A., & Leonard, L. B. (1978). Aphasics' comprehension of contextually conveyed meaning. *Brain and Language, 6,* 362–377.

Winner, E., Brownell, H., Blum, A., & Pincus, D. (1994). *Is an intact right hemisphere necessary for an intact theory of mind?* Manuscript submitted for publication.

Winner, E., & Gardner, H. (1977). The comprehension of metaphor in brain-damaged patients. *Brain, 100,* 717–729.

Zaidel, E. (1978). Lexical organization in the right hemisphere. In P. A. Buser & A. Rougel-Buser (Eds.), *Cerebral correlates of conscious experience* (pp. 177–197). Amsterdam: Elsevier/North Holland.

Zaidel, E. (1983). Disconnection syndrome as a model for laterality effects in the normal brain. In J. B. Hellige (Ed.), *Cerebral hemisphere asymmetry: Method, theory, and application* (pp. 95–151). New York: Praeger.

METAPHOR AND SYMBOLIC ACTIVITY *11*(1), 85–100
Copyright © 1996, Lawrence Erlbaum Associates, Inc.

Computational Approaches to Figurative Language

James H. Martin
University of Colorado

Computational approaches to figurative language are concerned with those situations in which the appropriate conceptual representation of an utterance differs fundamentally from that which could be composed from the ordinary meanings of the utterance's constituent words. The bulk of computational work in this area has addressed idiom, metonymy, metaphor, and indirect speech acts. These phenomena are generally recognized as being major impediments to the construction of successful natural language processing systems. Most such systems have attempted to deal with such figurative language through the use of extended notions of language conventions, general purpose reasoning methods, and strong notions of contextual expectations.

Computational approaches to figurative language are concerned with those situations in which the appropriate conceptual representation of an utterance differs fundamentally from that which could be composed from the ordinary meanings of the utterance's constituent words. Although there are a large number of figurative language devices that lead to such situations, the bulk of the computational work in this area has addressed idiom, metonymy, metaphor, and indirect speech acts. These phenomena are generally recognized as major impediments to the construction of successful natural language processing systems.

Such language is, of course, ubiquitous and, for the most part, easily handled by people. The question for those engaged in computational modeling is, if the standard resources and techniques for composing representations are inadequate for figurative language, then what other resources and techniques should be brought to bear? Computational researchers have, for the most part, drawn inspiration for their models from various schools of research in the cognitive sciences. Broadly speaking, these approaches have

Requests for reprints should be sent to James H. Martin, Computer Science Department and Institute of Cognitive Science, University of Colorado, Boulder, CO 80309–0430.

fallen into three categories: convention-based approaches, reasoning-based approaches, and context-based approaches.

Convention-based approaches rely on some explicitly enumerated set of facts that allow a language processor to associate one type of representation with another. Such approaches can be seen as extensions of the traditional Saussurean form-meaning model. In the case of idiom, the extension allows the association of multiword forms with specific meanings possibly unrelated to the meanings of the individual words. In the case of metaphor, metonymy, and speech acts, the extension has been more radical: the direct structured association of specific concepts with other concepts.

Reasoning-based approaches eschew the notion of figurative language conventions altogether. Rather, these approaches focus on what can be computed from the conceptual representation formed compositionally from the utterance. That is, there is assumed to be some procedure by which a strictly compositional meaning is taken as input and a new more appropriate representation is produced as output. Work in this vein often takes the view that figurative language processing is best approached as a problem for a general cognitive processing faculty rather than as a specifically language-related phenomenon. It follows, therefore, that the computational models following this approach are often based on preexisting models intended for more general reasoning tasks.

The bulk of the work in this vein has been on metaphor and indirect speech acts. In the case of metaphor, the reasoning is typically accomplished by some form of analogical reasoning system. Speech act processing is normally accomplished through the use of a general purpose rule-based reasoning system that uses the participants' shared knowledge of their mutual plans and goals.

Finally, by context-based approaches, I mean those approaches in which context has the primary role of proposing plausible meanings rather than merely selecting among meanings generated by other methods. Such approaches assume that figurative uses occur within a richly structured current context. This context typically includes a set of concepts created, or activated, by the previous discourse. This extended notion of context is usually assumed to include a likely meaning for a subsequent figurative utterance. In these approaches, the actual sentence "meaning" of an utterance is best seen as a cue to be used to select and possibly flesh out some part of a current discourse representation. The inspiration for this work tends to come from two areas of research in cognitive psychology: studies concerning the time needed to comprehend figurative language in context and research on discourse processing that focuses on how people use inference to construct conceptual representations of text.

The remaining sections of this article present overviews of various computational approaches to idiom, metonymy, metaphor, and indirect speech acts organized along the lines of these three general approaches. Before describing these efforts, I will briefly sketch in the next section what is usually assumed to be involved in the garden-variety semantic interpretation of literal language in natural language processing research. I should note that neither this description nor any of the descriptions of actual figurative language systems will be at a particularly technical level. Readers interested in the details of computational semantic analysis should consult Allen (1995), Hirst (1987), and Gazdar and Mellish (1989). Recent collections addressing the current technical aspects of figurative language processing can be found in Fass, Hinkelman, and Martin (1991, 1992), Barnden and Holyoak (1994), and Pustejovsky and Bergler (1992).

A BRIEF SKETCH OF COMPUTATIONAL SEMANTICS

In the vast majority of computational semantic research, the lexicon is taken to be a simple repository of mappings from forms to some kind of underlying meaning representations. In most models, the representation underlying nouns amounts to some type of attribute value list, often augmented with a type hierarchy. Verbs are typically treated as predicates with a fixed set of typed argument slots following some version of a deep case representation (Fillmore, 1968). The conceptual analysis of a single utterance consists of composing a representation from the representations associated with the individual constituent words according to composition rules either explicitly or implicitly associated with the syntactic structure of the sentence.

The principal problem that has to be addressed by such systems is controlling the considerable ambiguity that is inherent in the approach. This ambiguity arises from the large number of homonymous and polysemous word senses as well as from the grammar itself. Without some method of controlling this ambiguity, a natural language processing system would be faced with hundreds of representations for a single utterance. The most popular methods of dealing with ambiguity use a combination of semantic selection restrictions on the arguments to predicates, real-world commonsense knowledge, and knowledge of the immediate discourse context. It is hoped that, when combined, these techniques will be sufficient to select a single appropriate representation from the range of possible ones that might be created compositionally.

Unfortunately, in the case of figurative language, this approach usually winds up eliminating all candidate meanings because the figurative nature of the input leads them to be ill-formed in most cases. Of course, because the filtering of ill-formed representations is what these systems rely on to control ambiguity, the pervasive use of figurative language poses a major theoretical and practical hurdle to the ultimate success of these systems.

IDIOM

Idioms are the simplest and earliest form of figurative language to be addressed by the computational community. Unfortunately, they have also proven to be difficult for a number of reasons, but mainly because most idioms do not adhere to the simple notion of a fixed sequence of words associated with a fixed meaning. Instead, they display a fascinating array of syntactic and semantic variety that makes it difficult to come up with a unified computational framework capable of dealing with all types of idioms.

The bulk of computational work specifically addressing idiom has been driven by the need to account for this observed productivity (Becker, 1975; Chin, 1992; Jurafsky, 1993; Lytinen, Barridge, & Kirtner, 1992; Stock, 1989; van der Linden, 1992; Wilensky & Arens, 1980; Zernik, 1987; Zernik & Dyer, 1987). To illustrate some of the issues involved, consider the following examples:

> *"John and Mary tripped the light fantastic."*
> *"John and Mary buried the hatchet."*
> *The more I drink the sillier I get.*

The *"tripped the light fantastic"* use is a clear example of the simplest form of idiom: Its meaning seems to have no relation to any of the constituent parts, it resists any kind of syntactic or semantic productivity, and it is distinctly at odds with the normal constraints of English grammar. Idioms such as this correspond most closely to the simple notion of an essentially fixed and arbitrary sequence of words with a special meaning. Computationally, such idioms are dealt with by simply associating this meaning with this fixed sequence of parts and indexing the pattern under an appropriate grammatical category. In addition, in cases like this, there is no ambiguity problem because there is no valid literal interpretation.

Unfortunately, one does not have to look far to find example idioms that violate one or all of these typical characteristics. Consider the *"bury the hatchet"* example. The source of difficulty posed by this idiom is that there seems to be an intuition that the notions of "burying" and "hatchets" play some deep, if unclear, role in the meaning of this construction (Gibbs, 1989). This intuition becomes important when one considers the syntactic productivity of the construction. For the *"bury the hatchet"* idiom, this degree of compositionality seems to be enough to sanction the well-formedness of the passive version of the idiom. This is as opposed to the otherwise similar *"kick the bucket,"* where the bucket seems to play no underlying role and correspondingly exhibits no productivity.

Despite a somewhat bewildering array of different formalisms, nearly every currently implemented computational system that addresses this issue

(Chin, 1992; Jurafsky, 1993; Lytinen et al., 1992; Wilensky & Arens, 1980; Zernik & Dyer, 1987) makes the same claim. The syntactic productivity of a stored idiom arises from the association of parts of the surface pattern with parts of the underlying conceptual representation of the idiom's meaning. In some cases, as in Zernik and Dyer's RINA system, a further attempt is made to motivate the association between the part of the pattern and the underlying role. In this case, something akin to the notion that hatchets are weapons and burying them prevents their use would constitute such an explicit motivation.

Finally consider the *"the more I drink"* example. In their *construction grammar* approach, Fillmore, Kay, and O'Connor (1988) dubbed the underlying motivation for this use the "the Xer the Yer" construction. The only lexically specified parts of this idiomatic construction are the two uses of *the.* The remaining components are defined via a complex mixture of syntactic, semantic, and pragmatic parts. Current computational approaches to construction grammar (Jurafsky, 1993) are concerned with coming up with knowledge representation formalisms expressive enough to capture this complex mixing of constraints that, at the same time, still allow the implementation of efficient conceptual analysis programs.

In addition to these attempts to account for idiomatic generativity in a more principled way, recent work (Jurafsky, 1993; Lytinen et al., 1992; van der Linden, 1992) also began to focus on the psychological plausibility of the underlying representations and their associated processing models. This work is driven by the wealth of empirical timing data now available about human processing of idiomatic language (Cacciari & Tabossi, 1988; Gibbs & Nayak, 1989). These models attempt to accurately predict the initial time of activation of an idiomatic construction as well as the point at which the evidence is sufficient to select the idiom over competing analyses.

METONYMY

Metonymy has long been recognized as a problem for natural language processing systems. The obvious approach, not unlike similar approaches to idiom, is to simply list all possible senses of those words that are frequently used metonymically. For example, a word such as *newspaper* would simply have a large number of distinct senses to account for the following uses.

Today's newspaper is in my driveway.
Did you read today's newspaper?
Don't believe what you read in the newspaper.
Mary's father works for the newspaper.
The local newspaper is located downtown.

A lexicalist approach would simply list distinct senses for the physical copy of a single newspaper, an edition, a kind of medium, an organization that publishes a newspaper, and its location. This is somewhat problematic because it leads to an impractically large lexicon that, at the same time, is inadequate because there is no mechanism for dealing with unanticipated metonymic uses (i.e., a lexicon that is large and often does not work).

In addition to its practical shortcomings, this approach fails to capture any of the systematicity underlying most uses of metonymy. For example, the word *magazine* displays many, if not all, of the aforementioned senses of *newspaper*. Coming up with a way to exploit this type of systematicity is the motivation behind most computational approaches that directly address metonymy.

The META5 system (Fass, 1988, 1991) attempts to capture these systematicities through the use of a set of rules. These rules allow the system to specify that certain conceptual categories can be used to refer to other related categories. For example, to handle well known *"AUTHOR FOR WORK"* examples such as *"Plato is on the top shelf,* Fass's system simply encodes that schema as a rule. During interpretation, the mention of an author can then be used to produce a representation of the author's work. This new representation is placed into whatever conceptual role the author played in the utterance. To control its application, this mechanism is only triggered by the presence of some kind of semantic violation in the original utterance. Rules applicable to the concept causing the violation are then fired in an attempt to find another concept that alleviates the violation.

Pustejovsky's (1991) work on the *generative lexicon* represents another attempt to account for the systematicity underlying many metonymic uses. This work attempts to make the lexicon a more active generative device rather than a passive list of words. More specifically, the semantic representation of a noun in a given utterance is viewed as having been generated during the interpretation process rather than merely looked up. The generated representation is produced by combining information about the conceptual type of the noun with the constraints being imposed on it by the head of its verb phrase. In such an approach, metonymy is not an anomalous situation to be dealt with by a special mechanism. Rather, the interpretation of all nouns requires this type of sentence-specific sense generation. This approach is somewhat similar to Fass's (1988, 1991) approach in that it does, in effect, enumerate the functional roles that nouns may take on in combination with verbs.

Unfortunately, neither of these approaches are sufficient to deal with nonconventional metonymic uses such as the following.

"The ham sandwich didn't leave a tip."

It would be implausible for Fass's (1988, 1991) META5 system to have a preexisting *"ORDERED DISH STANDS FOR CUSTOMER"* rule in its knowledge base. Similarly, Pustejovsky's (1991) theory explicitly excludes this type of contextually motivated metonymy from the purview of the generative lexicon, leaving it to some more general discourse reasoning capability.

The Metallel system (Iverson & Helmreich, 1992), on the other hand, attempts to provide a mechanism that can deal directly with this type of metonymy. In fact, Iverson and Helmreich went further and argued that metonymic conventions of any type are unnecessary, because if some other more general reasoning mechanism is needed to deal with open-ended context-generated metonymys, that same general mechanism can deal with the more mundane examples as well. In the Metallel system, this is accomplished via a limited spreading activation technique. Spreading activation refers to a method whereby a system accesses concepts directly and indirectly related to those mentioned directly in a sentence.

Metallel uses spreading activation to implement an intersection search through its knowledge base of facts in an attempt to determine an appropriate referent for some noun phrase. In the ham sandwich case, the system would presumably find that the best connection in context from ham sandwich to tipping was through the customer concept, concluding that the customer who had ordered the ham sandwich was also doing the tipping.

As we will see with similar approaches to metaphor, this approach is prone to the overgeneration problem. Systems like Metallel exploit any of the available domain relations between the source and target concepts of the metonymy. However, the system falls short as an explanatory theory of metonymy because only a small fraction of such relations ever give rise to well-formed metonymies.

METAPHOR

Although metaphor has long been of concern to researchers in natural language processing (DeJong & Waltz, 1983; Russell, 1976; Wilks, 1975, 1978), much of current interest in the topic can be traced to the influence of Lakoff and Johnson's (1980) *Metaphors We Live By.*

Convention-Based Approaches to Metaphor

Taking direct, and sometimes overly literal, inspiration from this work, there are by now a large number of convention-based approaches to metaphor interpretation and generation (Hayes & Bayer, 1991; Jacobs, 1987; Jones & McCoy, 1992; J. Martin, 1990, 1992; Norvig, 1987, 1989; Veale & Keane, 1992). These approaches make the assumption that some core set of conceptual metaphors should be explicitly represented as part of a system's knowledge. To illustrate how such conventions can be exploited computationally, consider the following example.

"IBM got out of the typewriter business."

This example is based on a common conceptual metaphor that allows activities to be viewed as containers. In this instance, it makes use of the notion entailed by this metaphor that exiting an activity viewed as a container implies ceasing participation in the activity. More computationally, such a rule would explicitly link facts associated with physical containers with otherwise separate facts about activities.

When a concept is created compositionally that matches the source part of such a conventional metaphorical rule, the rule is used to create a new concept as specified by the target part of the rule. For this example, the semantic interpreter would produce a representation specifying something akin to a physical exiting of an physical object from a container. A concept representing IBM would fill the role of the physical object, whereas a representation of "the typewriter business" would play the role of the container. This skeletal representation would, in turn, activate an appropriate set of metaphorical association rules, which would, in turn, be used to produce a representation that replaces the notion of exiting with one capturing the notion of ceasing participation in some activity.

Of course, the various computational mechanisms embodying this approach instantiate these notions in a number of different ways. The primary differences involve the amount of structure that needs to be explicitly captured in the representations and the specificity of the encoded metaphors. These issues reveal the differing ways that computational researchers have interpreted the claims made by Lakoff and Johnson (1980).

One such view is that our conceptual system is structured by an extremely small number of abstract, loosely structured schemas. This view typically focuses on abstract orientational and body-centered metaphors. Veale and Keane (1992) coined the term *conceptual scaffolding* to refer to this view. Under this approach, a system would be expected to need an extremely small number of these abstract metaphorical schemas. During the process of interpretation, these schemas only provide the scaffolding for the final representation. The more detailed representations required for actual domain applications would be produced by augmenting these initial representations with long-term facts about the target domain and context specific inferences. Such an approach has the immediate practical advantage of requiring very little in the way of knowledge engineering. Natural language processing applications need only be provided with a small set of presumably universal metaphorical schemas to handle a wide range of metaphorical language.

At the other end of the spectrum is the work of J. Martin (1994) and Jones and McCoy (1992). Both of these efforts assume the need for a much larger number of highly structured metaphor schemas organized into some kind of hierarchical structure. Much of this work is driven by the desire to have the metaphorical knowledge base bear the primary burden of accounting for the wide range of metaphors observed in real texts.

In J. Martin's (1994) METABANK work, this desire arises directly from empirical research performing a metaphorical analysis of a large corpus. J. Martin claimed that an extremely small knowledge base of highly abstract orientational metaphors is simply not sufficient to account for the range of metaphors that are actually encountered in practice. In most genres of text, there appear to be a large number of apparently idiosyncratic domain-to-domain schemas. Although such metaphors have clear motivations based on more abstract metaphorical schemas, their existence in a given domain is just as clearly not predictable from the abstract schemas alone.

The work by Jones and McCoy (1992) followed a similar path from a more practical point of view. Jones and McCoy directly addressed the issue of generating metaphors from underlying semantic representations. Their concern was driven by the generative power of more abstract mappings. Simply put, they predicted a large number of metaphors that are never seen and would likely be judged anomalous by hearers. To ensure that their Quipper system only generates well-formed metaphors, Jones and McCoy made use of metaphors that are fairly specific and highly structured.

Reasoning-Based Approaches to Metaphor

Reasoning-based approaches to metaphor attempt to exploit information about the source and target concepts of a metaphor to determine an appropriate meaning without using prior knowledge of metaphorical conventions. The dominant paradigm in this area is the analogical approach (Carbonell, 1982; Fass, 1988, 1991; Gentner, 1983; Gentner, Falkenhainer, & Skorstad, 1988; Indurkhya, 1987; Weber, 1994). This approach asserts that metaphor depends on inherent structural similarities between the disparate source and target domains. It is the job of an analogical reasoning engine to uncover and specify the exact nature of the correspondences.

The simplest approaches address the comprehension of metaphors that take the form of *"X is a Y."* The source and target concepts are fully specified to the model. Following traditional views of metaphor, the task of the model is to use its knowledge of the source and target domains to alter the representation of the target concept. This process has typically involved one of three operations: changing the value of an existing feature in the target, changing the level of salience or prominence of a feature of the target, or transferring a feature–attribute pair from the source to the target (Chandler, 1991; Weiner, 1984).

Although theoretically interesting, this type of metaphor has an extremely low frequency in actual text. More critical to natural language processing applications are analogical approaches that directly address the more frequent relational, or predicate, kind of metaphor. Consider the previous container metaphor again:

"IBM got out of the typewriter business."

The first heuristic used in most systems is that the predicate-like term is interpreted as belonging to the source domain with some or all of its arguments belonging to the target domain. This heuristic seems to hold for a large number of examples and has some psycholinguistic support (Gentner et al., 1988) but is by no means inviolable, as can be seen in examples like *"John married a gem."*

The task of an analogical metaphor interpreter is to find or create the correct set of relations linking IBM to the typewriter business by making reference to concepts relating to getting out of containers. This is done without making use of any explicit associations between these domains.

This involves establishing a correspondence between the two domains. A significant part of this step is given by the predicate-argument structure of the original input. In our example, syntactic knowledge combined with case-role information about "get out of" tells us that IBM corresponds to the expected enterer, with the typewriter business playing the thing gotten into. Schematically, this information leaves us with the following analogical formula:

physob—exit—container::IBM—???—typewriter-business

The principle remaining task for the analogy engine is to find the missing concepts that will link up the specified pieces of the target concept in a way that most harmoniously satisfies the constraints imposed by this metaphorical equation. Of course, in any reasonably robust knowledge base there will be a large number of concepts that could relate the concepts of IBM and the typewriter business. This situation often leads to an exponentially large number of candidate analogical matchings between the source and target domains. The principal difference among the various analogical approaches to metaphor is the method they use to assess the well-formedness of candidate matchings.

The work by Gentner and colleagues (Gentner, 1983; Gentner et al., 1988) used her *systematicity principle* to judge the well-formedness of candidate solutions. This principle favors those solutions that preserve interrelated sets of relations among the various source and target concepts. In this scheme, this formalized notion of interrelatedness takes precedence over solutions based on features of individual objects or solutions based on matching individual relations. Indurkhya (1987) offers an alternative structural approach with different well-formedness criteria.

Fass (1991), in his work on META5, focused on the need to match different relations from the source and target domains. Unlike much of the work on analogy, Fass did not assume that the names of relations used to structure disparate domains will ever match exactly. Rather, relations, like other concepts, must be arranged into a hierarchy. META5 is based on a hierarchical matching scheme in which matches between relations are based on where in the abstraction hierarchy a common ancestor relation is found. Candidate solutions are then evaluated based on the degree of abstractness and interrelatedness of component relations in the differing solutions.

Carbonell (1982) took an approach that focuses on the content rather than on the structure of analogical solutions. This approach is based on the notion that certain relations should be privileged with respect to analogical matching. Carbonell suggested that a prioritized list of favored relations should be consulted when evaluating possible matches.

The primary tension between all of these approaches and approaches based on metaphorical conventions arises from the nature of the representation of many target domains, particularly those that seem to have an impoverished semantics absent some conventional metaphor. Consider the well-studied *"LIFE IS A JOURNEY"* metaphor. In the analogical-reasoning view, there must be a preexisting similarity between the conceptual representations of *"JOURNEYS"* and *"LIFE."* This similarity forms the basis for the metaphor and is discovered during the interpretation process. We can, however, ask whether it is not the case that our conception of the category *"LIFE"* is itself structured via the *"JOURNEY"* concept. If this is the case, then analogical approaches are attempting to rediscover the structural similarities between domains when one was created from the other as a template. Approaches based on convention take it to be more sensible to directly represent that history in the conceptual category structure than to rediscover it with each instance.

The most recent work by Indurkhya (1992) addressed this issue from a developmental viewpoint. The role of the analogy engine is shifted from the point of view of a competent processor of metaphors to the role of a developmental mechanism. In this view, the critical task for the analogy engine lies in the metaphorical creation of target concepts from well-specified source concepts. In his view, both the convention-based approach and the analogical-matching approach beg the question of how the target concept gets structured the way it does. The bulk of Indurkhya's computational modeling work has focused on the problem of metaphors with some visual basis.

Context-Based Approaches to Metaphor

As with most other issues in conceptual analysis, the context within which a metaphor is produced has a profound effect on its meaning. A number of researchers have made context the starting point for their theories of metaphor processing (Hobbs, 1979; C. Martin, 1993). They are supported by a wealth of psycholinguistic data showing significant processing differences for metaphors encountered in supporting contexts as opposed to those presented in isolation. The vast majority of these efforts attempt to show that the processing of metaphor can be made to fall out from mechanisms already needed to deal with nonmetaphorical text.

Note that these efforts do not merely make the assumption, common to nearly all computational models, that context is needed to select among plausible meanings proposed by some other context-independent semantic processor. Rather, they assert that context has the primary role of suggesting plausible representations. Under this framework, the semantic content of a new sentence is viewed as a set of clues that are to be used to find an appropriate way to extend the existing representation of the context. In the case of Hobbs (1979), this was accomplished by a limited context-directed inferencing system. C. Martin's (1993) direct memory access parsing (DMAP) system is based on an example-based approach that links metaphors in new utterances to specific remembered contexts of past use.

INDIRECT SPEECH ACTS

Work on speech act processing arises from the importance of discerning speaker's goals while they are interacting with various natural language processing applications. Consider the following example from a help system for the UNIX operating system:

> *"Can you tell me how to edit a file?"*
> *"Do you know how to delete a file?"*

The literal, or compositional, interpretations of these utterances clearly do not capture what the user is really trying to convey. As with other figurative approaches, computational research into this phenomenon has been divided into two camps: The first camp asserts that an appropriate meaning can be derived by reasoning about the domain semantics; the second asserts that specialized knowledge about the nature of specific speech acts must be employed.

Reasoning-based approaches to indirect requests have focused on the use of knowledge about the shared plans and goals of question askers and answerers (Allen & Perrault, 1980; Cohen & Perrault, 1979). In particular, users of such systems are assumed to have the goal of finding out how to do things with the system. More specifically, it can be assumed that they intend to acquire this knowledge by asking questions, the answers to which will provide the needed information.

Such approaches have been problematic for two reasons: They are quite costly in terms of the amount of time needed to compute what should be an easily derived meaning, and they suffer from the overgeneration problem. Specifically, they fail to tightly predict the restricted lexical and grammatical forms that are typically used with such speech acts.

Paralleling work on metaphor and metonymy, convention-based approaches to indirect requests come in two varieties: conceptual associations and surface-level constructions (Hinkelman, 1989; Hinkelman & Allen, 1989; Mayfield, 1992). Conceptual associations for speech acts are similar to those employed for conventional metaphors. Representations of the literal meanings are explicitly linked to the appropriate meanings via direct associations. For example, the construction underlying the previous examples would state that asking about someone's prerequisite ability to perform an action is a way of asking them to explain the action. This single rule accounts for both examples.

Of course, such an approach immediately leads to an ambiguity problem because both the literal meaning and the figurative meaning are possible. Both Mayfield (1992) and Hinkelman and Allen (1989) addressed computational ways to deal with this situation in the context of interactive natural language help systems.

Unfortunately, these conceptual conventions suffer from the usual overgeneration problem. Consider the following examples:

"Can you tell me how to delete a file?"
"?Are you able to tell me how to delete a file?"

The second example corresponds to many of the schemas suggested for handling the first example. It is, however, clearly ill-formed in some way and simply does not occur with the indirect meaning. Hinkelman's (1989) thesis work addressed this problem by adding a mix of grammatical and semantic constraints to the convention, thereby constraining the range of productivity of its use.

SUMMARY

The earliest computational work on figurative language can be characterized as an attempt to simply get something to work within some existing application framework. Getting something to work meant coming up with some scheme in which some representation judged appropriate to an application could be produced from what would otherwise have to be considered an ill-formed input. These schemes, when viewed from either a theoretical or practical perspective, were found wanting for several reasons including their failure to hold up when the issue of productivity or generativity was considered, their failure to match up well with known psycholinguistic constraints concerning reaction times, and their failure to adequately account for the influence of context. Contemporary computational systems have had considerable success in addressing many of these issues. There remains, however, a distinct lack of consensus in the field as a whole. There has, in particular, been little movement toward integrating successful techniques from the various approaches into a unifying framework.

It is important to note that these computational efforts have derived considerable benefit from paying close attention to empirical psycholinguistic constraints concerning timing data. Further progress may be made toward a unifying framework if other empirical constraints are taken seriously. Three sources for such constraints are ripe to be exploited: corpus-based data concerning naturally occurring figurative uses in context, cross-linguistic data becoming available through the efforts of those working on machine translation, and, finally, the considerable developmental data now available as a part of the Childes project (MacWhinney, 1991).

REFERENCES

Allen, J. (1995). *Natural language understanding* (2nd ed.). Redwood City, CA: Benjamin-Cummings.

Allen, J., & Perrault, C. (1980). Analyzing intention in utterances. *Artificial Intelligence, 15,* 143–178.

Barnden, J., & Holyoak, K. (Eds.). (1994). *Analogy, metaphor and reminding.* Norwood, NJ: Ablex.

Becker, J. (1975). *The phrasal lexicon.* Paper presented at the Workshop on Theoretical Issues in Natural Language Processing, Cambridge, MA.

Cacciari, C., & Tabossi, P. (1988). The comprehension of idioms. *Journal of Language and Memory, 27,* 668–683.

Carbonell, J. (1982). Metaphor: An inescapable phenomenon in natural language comprehension. In M. Ringle & W. Lehnert (Eds.). *Strategies for natural language processing* (pp. 415–434). Hillsdale, NJ: Lawrence Erlbaum Associates, Inc.

Chandler, S. (1991). Metaphor comprehension: A connectionist approach to implications for the mental lexicon. *Metaphor and Symbolic Activity, 6,* 227–258.

Chin, D. (1992). Pau: Parsing and understanding with uniform syntactic, semantic and idiomatic representations. *Computational Intelligence, 8,* 456–476.

Cohen, P. R., & Perrault, C. (1979). Elements of a plan-based theory of speech acts. *Cognitive Science, 3,* 177–212.

DeJong, G. F., & Waltz, D. L. (1983). Understanding novel language. *Computers and Mathematics with Applications, 9,* 131–147.

Fass, D. (1988). *Collative semantics: A semantics for natural language* (CRL Rep. No. MCCS–88–118). Lac Cruces: New Mexico State University.

Fass, D. (1991). met*: A method for discriminating metaphor and metonymy by computer. *Computational Linguistics 17,* 1, 49–90.

Fass, D., Hinkelman, E., & Martin, J. (Eds.). (1991). *Proceedings of the IJCAI Workshop on Compuational Approaches to Non-Literal Language: Metaphor, metonymy, idiom, speech acts and implicature.* (Tech. Rep. No. CU–CS–550–91). Boulder: University of Colorado, Computer Science Department.

Fass, D., Hinkelman, E., & Martin, J. (Eds). (1992). Non-literal language [Special issue]. *Computational Intelligence, 8,* 1–88.

Fillmore, C. (1968). The case for case. In E. Bach & R. T. Harms (Eds.), *Universals in linguistic theory* (pp. 1–88). Chicago: Holt, Rinehart & Winston.

Fillmore, C., Kay, P., & O'Connor, M. (1988). Regularity and idiomaticity in grammatical constructions: The case of let alone. *Language, 64,* 510–538.

Gazdar, G., & Mellish, C. (1989). *Natural language processing in LISP.* Reading, MA: Addison-Wesley.

Gentner, D. (1983). Structure mapping: A theoretical framework for analogy. *Cognitive Science, 7,* 155–170.

Gentner, D., Falkenhainer, B., & Skorstad, J. (1988). Viewing metaphor as analogy. In D. Helman (Ed.), *Analogical reasoning* (pp. 171–177). Norwell, MA: Kluwer Academic.

Gibbs, R. (1989). Understanding and literal meaning. *Cognitive Science,* 13, 243–251.

Gibbs, R., & Nayak, N. (1989). Psycholinguistic studies on the syntactic behavior of idioms. *Cognitive Psychology, 21* (13), 100–138.

Hayes, E., & Bayer, S. (1991). Metaphoric generalization through sort coercion. In *Proceedings of the 29th Meeting of the Association for Computational Linguistics* (pp. 222–228). Morristown, NJ: Association for Computational Linguistics.

Hinkelman, E. (1989). *Linguistic and pragmatic constraints on utterance interpretation* (Rep. No. 288). Rochester, NY: University of Rochester, Computer Science Department.

Hinkelman, E., & Allen, J. (1989). Two constraints on speech act ambiguity. In *Proceedings of the 27th Annual Meeting of the Association for Computational Linguistics* (pp. 212–219). Morristown, NJ: Association for Computational Linguistics.

Hirst, G. (1987). *Semantic interpretation and the resolution of ambiguity.* Cambridge, England. Cambridge University Press.

Hobbs, J. (1979). *Metaphor, metaphor schemata, and selective inferencing* (Tech. Rep. Tech. Note No. 204). San Mateo, CA: SRI.

Indurkhya, B. (1987). Approximate semantic transference: A computational theory of metaphors and analogy. *Cognitive Science, 11,* 445–480.

Indurkhya, B. (1992). *Metaphor and Cognition.* Dordrecht, The Netherlands: Kluwer.

Iverson, E., & Helmreich, S. (1992). Metallel: An integrated approach to non-literal phrase interpretation. *Computational Intelligence, 8,* 477–493.

Jacobs, P. S. (1987). Knowledge-intensive natural language generation. *Artificial Intelligence, 33,* 325–378.

Jones, M. A., & McCoy, K. (1992). Transparently-motivated metaphor generation. In R. Dale, E. Hovy, D. Rösner, & O. Stock (Eds.), *Lecture notes in artificial intelligence: Vol. 587. Aspects of automated natural language generation* (pp. 183–198). Berlin: Springer-Verlag.

Jurafsky, D. (1993). *A cognitive model of sentence interpretation: The construction grammar approach* (Tech. Rep. No. TR 93–077). Berkeley, CA: Jurafsky.

Lakoff, G., & Johnson, M. (1980). *Metaphors we live by.* Chicago: University of Chicago Press.

Lytinen, S., Burridge, R., & Kirtner, J. D. (1992). The role of literal meaning in the comprehension of non-literal constructions. *Computational Intelligence, 8,* 416–432.

MacWhinney, B. (1991). *The CHILDES Project: Tools for analyzing talk.* Hillsdale, NJ: Lawrence Erlbaum Associates, Inc.

Martin, C. (1993). *Direct memory access parsing* (Tech. Rep. CS 93–07). Chicago: University of Chicago, Department of Computer Science.

Martin, J. H. (1990). *A computational model of metaphor interpretation.* Cambridge, MA: Academic.

Martin, J. H. (1992). Computer understanding of conventional metaphoric language. *Cognitive Science, 16,* 233–270.

Martin, J. H. (1994). Metabank: A knowledge-base of metaphoric language conventions. *Computational Intelligence, 10,* 134–149.

Mayfield, J. (1992). Controlling inference in plan recognition. *User Modeling and User Adapted Interaction, 2,* 55–82.

Norvig, P. (1987). *A unified theory of inference for text understanding* (Rep. No. UCB/CSD 87–339). Berkeley: University of California, Computer Science Department.

Norvig, P. (1989). Marker passing as a weak method for text inferencing. *Cognitive Science, 13,* 569–620.

Pustejovsky, J. (1991). The generative lexicon. *Computational Linguistics, 17,* 409–441.

Pustejovsky, J., & Bergler, S. (Eds.). (1992). *Lecture notes in artificial intelligence: Vol. 627. Lexical semantics and knowledge representation.* Berlin: Springer-Verlag.

Russell, S. W. (1976). Computer understanding of metaphorically used verbs. *American Journal of Computational Linguistics, 2,* 15–28.

Stock, O. (1989). Parsing with flexibility, dynamic strategies and idioms in mind. *Computational Linguistics, 15,* 1–18.

van der Linden, E.-J. (1992). Idioms, non-literal language and knowledge representation. *Computational Intelligence, 8,* 433–455.

Veale, T., & Keane, M. T. (1992). Conceptual scaffolding: A spatially founded meaning representation for metaphor comprehension. *Computational Intelligence, 8,* 494–519.

Weber, S. (1994). A structured connectionist model of figurative adjective–noun combinations. In J. Barnden & K. Holyoak (Eds.), *Analogy, metaphor and reminding* (pp. 45–59). Norwood, NJ: Ablex.

Weiner, J. E. (1984). A knowledge representation approach to understanding metaphors. *Computational Linguistics, 10,* 1–14.

Wilensky, R., & Arens, Y. (1980). Phran: A knowledge-based natural language understander. In *Proceedings of the Association for Computational Linguistics* (pp. 117–121). Philadelphia.

Wilks, Y. (1975). Preference semantics. In E. Keenan (Ed.), *The formal semantics of natural language* (pp. 329–348). Cambridge, England: Cambridge University Press.

Wilks, Y. (1978). Making preferences more active. *Artificial Intelligence, 11,* 197–223.

Zernik, U. (1987). *Strategies in language acquisition: Learning phrases from examples in context* (UCLA–TR No. UCLA-AI–87–1). Los Angeles: University of California.

Zernik, U., & Dyer, M. (1987). The self-extending phrasal lexicon. *Computational Linguistics, 13,* 308–327.